Praise for this book...

'"How do we apply PRA for animals?" was the rather strange query from a participant of Praxis's international workshop. That was the beginning of a journey of exploration of the use of participatory tools for the cause of animal welfare, and Praxis has been fortunate to be associated with that journey. *Sharing the Load* is commendable for its innovation and its deep commitment to participation. It is a must-read for all those involved not only in animal welfare, but also social development in general.'

Mr Tom Thomas, Chief Executive, Praxis, Institute for Participatory Practices, India

'This long-awaited book will help practitioners and animal welfare agencies improve the effectiveness of their operations both with working animals and with the people who own or work them. The authors have combined advanced knowledge of animal welfare (Pritchard and Wells) and community-based participatory methods (van Dijk and Pradhan) to produce a beautifully accessible and practical book. An essential guide for anyone providing development assistance where there are working equines, and applicable to other working animal species.'

David Hadrill, veterinary consultant, member of the Board of Directors Vetwork UK

'This is a pioneering work of its own kind, which I am sure will contribute directly in improving the livelihoods and well-being of millions of poor people in Asia, Africa and Latin America who depend primarily or partly on income from working animals. Extensive use of participatory tools with visuals for a better and easier understanding of local situations make the manual more user-friendly, appropriate and attractive.'

Dr Kamal Kar, Chairman, CLTS Foundation, Kolkata, India

'This charming book conveys a great deal of knowledge and compassion for working animals in a most accessible form. It does not preach. Readers are first set free to develop their own understanding of animal welfare and the mutual dependence of working animals and their owners. They are then given a practical toolbox for use in the field with communities, however little or large their literacy. A lot of people and a lot of animals are going to feel better as a result of this book.'

John Webster, Emeritus Professor of Animal Husbandry, University of Bristol, UK

'*Sharing the Load* is a unique and comprehensive field guide for community facilitators working on animal welfare. The language used is simple and readable, and the illustrations are attractive. A must for all working on animal welfare and community participation.'

Somesh Kumar, Indian Administrative Service

'The important field experience of the Brooke gives an immense value to this guide and makes it a unique read. It is rich in clear and motivating cases that I found fascinating to read. Feedback from real life and practical examples are precious and I would have been happy to read them years ago when I was a practitioner trying to persuade farmers to invest their efforts in caring about the welfare of their animals.'

Dr Andrea Gavinelli, Head of Animal Welfare, Health and Consumers Directorate General, European Commission

Sharing the Load

A guide to improving the welfare of working animals through collective action

Lisa van Dijk, Joy Pritchard, S.K. Pradhan and Kimberly Wells

PRACTICAL ACTION
Publishing

Practical Action Publishing Ltd
Schumacher Centre for Technology and Development
Bourton on Dunsmore, Rugby,
Warwickshire, CV23 9QZ, UK
www.practicalactionpublishing.org

ISBN 978 1 85339 719 6

Since 1974, Practical Action Publishing (formerly Intermediate Technology Publications
and ITDG Publishing) has published and disseminated books and information in support
of international development work throughout the world. Practical Action Publishing is a
trading name of Practical Action Publishing Ltd (Company Reg. No. 1159018), the wholly
owned publishing company of Practical Action. Practical Action Publishing trades only in
support of its parent charity objectives and any profits are covenanted back to Practical
Action (Charity Reg. No. 247257, Group VAT Registration No. 880 9924 76).

Cover image: © Martha Hardy@GCI
Cover design: Practical Action Publishing
Illustrations: © Martha Hardy@GCI (Parts 1 and 2) and Amitabh Pandey (Part 3)
Typeset by S.J.I. Services, New Delhi
Printed by Replika Press Pvt. Ltd

Contents

PART III Participatory action tools for animal welfare

Figures

Tables

Case studies

Boxes

Theory boxes

Process boxes

Preface

Why did we write this manual?

Many people have knowledge of community facilitation and collective action. Many others have knowledge of working animals and their welfare. For the first time we bring together both areas of knowledge in a field manual for community facilitators. The manual describes the practical steps and tools needed to stimulate collective action for long-term, sustainable improvement in the welfare of working animals.

The tools in this manual have been developed and tested with owners of working horses, mules and donkeys. These tools are in daily use with animal-owning communities. We believe that they are useful for improving the welfare of other species of traction and transport animal (such as bullocks, buffalo, camels and yaks) because their working conditions and the livelihoods of their owners are similar to those of working horses and donkeys. We also believe that they can be adapted for improvement of the welfare of farm livestock and hope that some readers are motivated to develop them further for this purpose.

Who can use this manual?

This manual is written for community facilitators and anyone else who has direct contact with working (traction and transport) animals and their owners, including vets, community-based animal health workers, government extension workers and development workers.

This manual assumes that you, the community facilitator, already have the following skills, knowledge and attitudes:

- Basic skills in community mobilization, including building rapport, organizing people, listening skills, understanding how community groups work, and understanding the local language and conditions.

- Participatory rural appraisal (PRA) skills or similar training in the use of tools for stimulating community action for any kind of benefit.

- A desire for community empowerment and ownership of the project.

- Compassion for animals and a calm attitude towards them.

- A basic knowledge of animal care and husbandry is useful, but not essential.

How to use this manual

This manual has three parts:

- Working animals and their welfare

- Interventions for lasting change

- Toolkit for facilitators

The manual can be used in different ways.

If you are a community facilitator working with animals for the first time, we recommend that you read the book all the way through. You can then use the steps and tools to design and facilitate your programme, referring back to them regularly and building on them with your own ideas as your experience grows.

If you are an experienced community facilitator with an agricultural or animal health background, we suggest that you read Part 1 to become familiar with considering animal welfare alongside productivity and the needs of people. This part also describes the important differences between working animals and livestock. Then have a look at the Toolkit (Part 3) and incorporate some specialist animal welfare tools into your existing activities.

If you are a vet, animal health worker or extension worker who wishes to prevent disease or poor health through sustainable improvement in animal husbandry and management of working practices, we suggest that you skim the manual to familiarise yourself with the main headings and things that look most relevant to your work. Then read these chapters in depth. If you do not yet have the skills in community facilitation described above, you may wish to find further training or advice before starting to carry out a community programme.

Expectations when using this manual

Sharing the Load was written by a group of experienced community facilitators, vets and animal welfare scientists who have been working in this field for several years, making mistakes, learning from them and continuing to grow new ideas all the time. It has been developed with many groups of people who use working animals as a major part of their livelihood system. The communities who tested and adapted the tools have been a vital part of the process and taught us a lot about what is and is not useful for them.

Using this manual does not guarantee success in stimulating collective action for lasting improvement in animal welfare. This will require time, skill, experience and perseverance in working with communities in a participatory way, blending in the new knowledge found in this manual and adapting to the feelings, needs and wishes of each community and their animals. It also requires a 'climate for change' to be created within the community, the facilitator and the facilitating or funding organization. Further information can be found in the chapter 'Interventions for lasting change'.

This book is not suitable for use in disaster or conflict situations, where the time and conditions needed for effective community participation are not available. See the reference list at the back of the manual for further information on improving animal welfare in emergencies.

Most importantly, success relies on commitment to long-term improvement in the welfare of working animals because they are important for themselves and for the communities that depend on them.

Acknowledgements

We would like to thank all the owners, users and carers of working horses, mules and donkeys whose experiences have formed the basis for this book. Many thanks go to the Brooke's facilitators and those from our partner organizations worldwide, for their contributions arising from daily sharing and learning with animal-owning communities. In particular Ramesh Ranjan, Murad Ali and the Brooke India community development team have ensured that this manual reflects their real experiences with animal welfare facilitation in the field.

We are grateful to Anindo Banerjee, Tania Dennison and Helen (Becky) Whay whose participation in the initial 'writeshop' for this book is much appreciated. We would like to thank the Brooke UK staff and supporters who funded its production and Dorcas Pratt, Director of International Development, for her comments on the final draft.

Thanks to Martha Hardy from Graham-Cameron Illustration and to Amitabh Pandey for their beautiful illustrations which bring our writing to life.

Finally we would like to thank working animals all over the world for their invaluable contribution to the daily livelihoods of so many people and communities. We hope that this book is one more step towards making their lives better.

Symbols used in the text

	Theory box In a theory box you will find some theoretical background about the information discussed in the paragraph or chapter. Often there is a reference to further reading for a deeper understanding of the theory. The further reading and reference section is at the back of the manual.
	Process box A process box presents further explanation of a specific issue mentioned in the text, or summarises the key points or steps in a process described in the text. This will help you to use the process with the community.
	Case study box In a case study box you will find a real life example of the processes described in the text. These case studies provide you with an insight into how facilitators have used the process to mobilize animal owners to take action to improve the welfare of their working animals.
	Warning box There is one warning box in the book, in Chapter 4. If you see it, please take note of it!

Acronyms

AAAAQ	Accessibility-Availability-Affordability-Acceptability-Quality
APA	Appreciative planning and action
CBAHW	Community-based animal health workers
PLA	Participatory learning and action
PRA	Participatory rural appraisal
PWNA	Participatory welfare needs assessment
SHGs	Self-help groups

Introduction

Sharing the Load aims to improve the lives of the millions of animals that work – pulling carts, ploughing fields and carrying loads on their backs – by stimulating collective action among animal-owning communities to improve their animals' welfare. In doing so, people's livelihoods and relationships with their working animals may also be improved. Essential to this process is a community facilitator (motivator, extension worker, promoter or change agent) who can help create the right conditions for the community to act.

People working in the international development sector have devised processes, methods and tools to facilitate community development and action for change in healthcare, water and sanitation, agriculture and many other areas. In the course of our own work to improve the welfare of working animals, we identified a gap in the availability of field-based tools and methods for understanding and creating change in animal welfare.

Changing human behaviour is particularly challenging when the benefit is for a third party (the animal) rather than a person and his or her immediate family. There may be short-term costs in terms of effort, time, money and productivity in order to gain longer-term improvements for animals and the community. In addition, it is difficult to know how animals would describe their own welfare, if indeed they were able to tell us, and what they would choose to do to improve their welfare. Facilitators working with animal-owning communities can find it hard to adapt the existing tools and processes they know well, to a situation with which they are less familiar. This is why the idea for this manual was conceived.

Working animals and their welfare

Achieving good welfare, or well-being, for working animals can be considered more complex than for any other use of animals by people. As with farmed species, they are completely dependent on their owners and carers for food, shelter and other aspects of good husbandry. Unlike cattle, sheep, pigs and poultry, however, they tend to be kept and worked individually or in small groups with restricted opportunities for the social interaction and comfort gained from others of their own type.

Traction and transport animals experience more varied environments than any other animal, from carrying packs up the Himalayan foothills where no motorized transport can reach, to ploughing, harrowing, seeding and weeding farms in Africa, to moving every imaginable type of goods through the crowded and polluted streets of the world's biggest cities. These animals draw and distribute water, drag sand from rivers, pull minerals from mines, transport bricks and metal for building, thresh corn, take goods to market, move tourists and refugees, carry the sick to hospital and are often a vital part of weddings and ceremonial occasions. To shift their infinitely varying loads, working animals suffer many kinds of physical and mental stress. They may be subjected to extreme heat or cold, or to wet or dry climates; they may have to carry heavy and awkward burdens or walk over difficult terrain. They share the same hardships in life as their owners.

Interdependence between working animals and their owners: benefits and dilemmas for welfare

There is a strong bond of interdependence between a working animal and those who depend on it for a livelihood. People all over the world tell stories of their animals' importance and value. In these cases, it may be expected that a working horse, donkey, bullock or camel is always well-cared for and appreciated by those who interact with it and by society as a whole. Unfortunately, this is not necessarily the case.

Animal-owning communities are frequently constrained by factors such as poverty, low-status and restricted access to resources for their families or their animals. Compared with other sections of society they have few opportunities to improve their lives and will naturally prioritize their own health or their children's education, for example, above the well-being of their animals. Even among animal species, working animals are often the last to benefit from extra food or other resources that may become available, because their productivity (measured as work output or traction energy) is not recognized when compared to directly marketable products such as meat or milk.

To address these challenges, this manual suggests tools and methods to increase collective interest in working animals among their owners and communities and to translate this interest into action for improved welfare. As a result, it is hoped that people will choose the best course of action for themselves and make their own decisions about where, when and how to act. Along the way, you, as a facilitator, will witness heated discussions and even some resistance to focusing on the needs and feelings of animals when their owners face so many other problems. The changes that are agreed are likely to be small steps, rather than great leaps forward, but in our work all over the world, we have found that every community engages in the challenge.

Collective action for animal welfare improvement

What is collective action?

Collective action means working as a group, rather than as individuals, to achieve a common goal. In this case, the common goal is to improve the welfare of working animals.

Why is collective action essential to bring about change in animal welfare?

Experience from the human development sector shows that two important conditions are needed to bring ownership of any action by the community and enable the action to continue after the facilitation process is over. These are:

- A high level of motivation and enthusiasm for the action (animal welfare improvement) within the community.
- Effective community organization which can support and maintain the process and take it forward into the long-term future.
- Without these, there is little chance that welfare-promoting activities will be sustained by a community without continuous external support. Externally-motivated projects often fail once support either diminishes or is withdrawn.
- The basis for successful action is participation, so it is essential to create conditions for this right from the start.

Benefits of community-led participatory approaches

There are many benefits to community-led participatory approaches.

Knowledge

- International experience of participatory projects shows that local knowledge and wisdom are crucial for successful development processes.

- Participatory learning approaches encourage, support and strengthen a community's ability to identify their own animals' needs, set their own objectives for improving welfare, and to plan, implement, monitor and evaluate their own initiatives.

- Participatory learning approaches provide local communities with invaluable experience and skills for animal husbandry and handling, which are shared among everyone.

- Traditional wisdom which may be harmful, is questioned, while that which benefits animals is preserved or revived.

- People are encouraged to look for examples, within their communities, of approaches that have dealt successfully with similar problems in different ways.

Self-reliance

- Participation helps achieve self-reliance by breaking the dependency which has become inbuilt into many social systems and reinforced by free services or heavily subsidized external inputs from supporting agencies or charities.

- Community participation promotes awareness of animal welfare issues and the confidence to change.

- Participation can create and enhance a force which binds the community together and encourages acceptance of joint responsibility for their animals.

- Marginalized or deprived groups of people build their own capacity to examine their animals' problems and seek solutions collectively.

Effectiveness and sustainability

- Community participation makes it more likely that resources will be used efficiently and that processes will be effective.

- Participation enables the programme to expand its scope and incorporate a broader range of ideas than was first thought.

- Sustainability is built as the community takes on the existing activities of external agencies. It also develops local management systems to maintain these activities once external inputs are withdrawn.

- A participatory approach promotes collaboration within and outside the community, through networking activities or small animal-owning groups forming into federations, for example. This increases the likelihood of bringing about change and maintaining it.

How is collective action initiated and maintained?

Animal welfare is an unfamiliar issue for the owners, handlers and carers of working animals and initially it may be difficult to motivate them to focus on this topic. They are likely to be carrying out many good welfare practices in their daily animal husbandry, but may not recognize these as 'welfare' – a word which does not translate directly into most languages. As a facilitator, your first priorities will be to identify the key concerns about the welfare of working animals which the community would like to address and to unite them around a common activity or goal ('entry point'). People will only be motivated to take action when they identify such issues themselves and then discuss and formulate them into clearly-expressed needs for both animals and people, along with a common vision of the expected improved situation.

Following an initial discussion around common topics of interest, the participatory learning and action (PLA) process follows a series of phases or steps. These are not a fixed prescription; they are guidelines which can be adapted according to the local situation and your relationship with the community and the way it has responded to the challenge. Do not be put off by the sequence of steps described in this manual – they do not necessarily mean lots of elaborate planning and implementation. The important thing is that the community's wishes should come first and your own agenda for improved animal welfare should come after – it will be strengthened in the later stages of the process.

Owners of working animals have variable livelihoods. Different members of the family or community are often involved in managing animals whilst working and at home. If the experiences or interests of owners, handlers and carers are very mixed, they may not form a strong group. In this case, forming smaller groups of similar people (such as the wives of animal owners or the boys who hire donkey carts) may be more effective than larger mixed groups. These smaller groups can then decide how to associate and network with others to form a larger organization if and when they wish to manage broader issues of common interest.

Community groups will evolve their own rules, regulations and systems of management. These can lead to effective collective action within their local environment and will enable a group to sustain animal welfare interventions over long periods of time. A well-organized group will continue to function even after the withdrawal of your supporting agency and will provide a strong, stable institutional base from which to meet the requirements of the community and its animals.

Empowering animal-owning communities

In the context of this manual, empowering people means enabling the community (animal-owning families and other local stakeholders) to understand the reality of their present situation, to reflect on the factors leading to poor welfare of their working animals and to take steps to improve the situation. The community decides where it is now, where it wants to go, and makes a plan to reach these goals, based on self-reliance and sharing of power. Most importantly, it breaks the mind-set of dependence on others, so the community acquires the ability to decide its own path.

Empowered local groups will analyse their problems and think of possible solutions according to their own knowledge and understanding. They will examine alternatives that may be suggested by you or your supporting agency, consider the options and then decide what is most appropriate for them.

Your role as a facilitator is to introduce various participatory tools and methodologies to help the community to identify the issues, prioritize them and then discuss and act on the factors responsible for these issues. Special skills needed by a group facilitator or promoter include a clear understanding of the participatory approach, the ability to organize people and a willingness to listen and co-operate closely with others in partnership.

Theory box 1. Some challenges of empowerment

- Animal-owning communities have many problems of their own. A supporting agency's 'mission', 'programme', or 'target' is not necessarily seen as the most important thing or even as particularly relevant to them.
- Participation is not a one-off activity or input into projects. It is a process which evolves over time and whose direction and outcome are not always predictable or manageable.
- Participatory learning approaches enable the community to express their understanding or concerns about problems they consider to be important. The principle is to move the analysis from a problem towards possible solutions. This may sound simple but it can be very challenging to field staff from a supporting agency, who have to match their agency's mission and targets with the often very different concerns of the community.
- As a facilitator, you should bring in your own knowledge or that of an external agency only after the local community and groups have considered various options to deal with the problems that they consider important and have requested supplementary information.
- Action is only worth considering when the benefits of a solution become substantially more than the costs involved. This applies to any type of effort towards self reliance, including sustainable improvement in the welfare of working animals. Therefore it is important to try to ensure that the benefit-to-cost ratio is positive.

Theory box 2. Motivation to improve animal welfare

Animal welfare scientists and practitioners are constantly seeking new ways to motivate animal owners and carers to improve welfare. Different methods of motivation are appropriate for the various circumstances in which animals are kept and used. Here we illustrate some of the differences between externally-generated motivation, such as that found in many animal welfare training, accreditation and prosecution schemes, and the motivation generated internally within groups of animal owners working together to improve welfare.

Externally-generated motivation to improve animal welfare	Equivalent motivation generated internally by community participatory approaches
Education Animal welfare knowledge is introduced to owners from the outside through teaching, training or reading published materials.	*Sharing knowledge* Animal owners sit together to share and develop their existing local knowledge through discussion and joint analysis of problems.
Encouragement Trainers and supporting agencies give encouragement and reward for welfare improvement. Incentives may be created with external inputs such as farm subsidies, animal welfare benchmarking, free veterinary treatment, or paying a premium for animal products produced in a welfare-friendly way.	*Peer motivation* Members of the group encourage each other to attend regular meetings and participate in activities which will help improve animal welfare. A positive competitive spirit develops within the group, with each owner trying to get their animal into a better condition than the others.
Enforcement Quality assurance certification or accreditation schemes exclude farmers whose animals suffer poor welfare, so income from contracts or premium payments is lost, for example. In extreme cases, national or local legislation is used to prosecute animal owners and they are fined or imprisoned as a punishment.	*Peer pressure* Members of the animal-owning group put social pressure on each other to comply with the agreed rules of the group and to implement action plans for welfare improvement. Fines and penalties for breaking rules or norms are agreed collectively and followed up by the group.

Source: van Dijk, L. and Pritchard, J.C. (2010)

For further reading on Education, Encouragement and Enforcement: Main, D.C.J., Kent, J.P., Wemelsfelder, F., Ofner, E., and Tuyttens, F.A.M. (2003) 'Applications of methods of on-farm welfare assessment', *Animal Welfare* 12, 523–528

PART I
WORKING ANIMALS AND THEIR WELFARE

What you will find in Part I

This part contains an introduction to working animals. In Chapter 1 we look at the animals at work and their relationship of mutual dependency with people, exploring:

- what working animals are.
- why people have working animals.
- how working animals and their owners, families and communities depend on each other.

In Chapter 2 we look at what animal welfare means: the needs and feelings of working animals. We consider the behaviour used by animals to express their needs and feelings, and how to observe and interpret this behaviour in order to hear the 'voice of the animal'.
We also discuss:

- why good welfare is important for working animals;
- what determines their welfare state;
- what can be changed in order to improve their welfare.

How to use Part I

The chapters in this part of the manual are organized as a learning experience. You will need to read them in a place where you can walk around outside and look at animals as they are working. Ideally find a comfortable space, such as a tea stall near a market place where animals are gathered together, or on a street where they pass by regularly. You will then be able to read and do the short written exercises, and also walk around to observe the animals nearby. Each section is laid out in a similar way:

1. A short exercise or a few questions to answer in the spaces provided (or on a separate sheet of paper if you are sharing this book with others). These will stimulate your thinking and draw from your past experience. Some exercises will ask you to walk around, look at animals and make notes of your thoughts and observations.

2. Most exercises then have a picture which illustrates some things to consider in your answers.

3. A statement answers the question and helps you to test your knowledge. You may wish to make extra notes to go with the statement, based on your own experience and observations.

After you have read Part I, you should be able to think about the questions and exercises when you are out in the field and see animals at work. Look at the ways in which they are used by people, how people and their working animals interact with each other and how the interactions affect the animals' welfare. This will help to expand your experience and generate new questions or discussion points when you come to work with the community.

CHAPTER 1
Working animals and the communities who own them

In Chapter 1 you will become familiar with what working animals are, why people have them and how working animals and people depend on each other.

After reading this chapter you will be able to:

- identify what working animals are and what they do;

- recognize how people and animals are connected through their livelihoods and working systems;

- describe why people have responsibilities towards their working animals and what these responsibilities are.

What are working animals and what do they do?

Question 1

Start by looking at the animals around you in the village, market or street. Write down what they are doing. Which ones are working animals and which are not? What do we mean by 'working'?

....
....
....
....

How do working animals contribute to people's lives? How does this differ from the ways in which other livestock benefits people?

....
....
....
....

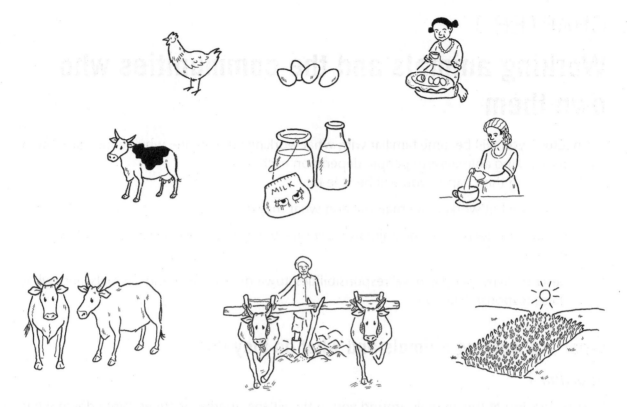

Livestock such as cows, sheep, goats and poultry produce food or fibre to support a livelihood such as milk, meat, eggs or wool. Working animals such as horses, donkeys, camels, bullocks or yaks pull or carry things or people using their energy and body power, by, for example, ploughing and harrowing fields or transporting people and goods. We refer to this use of their body energy for power as their productivity.

Question 2

Look again at the animals nearby and think about other working animals you have seen. What are they pulling and carrying? Ask the people working with them if their animals ever pull or carry anything else.

Pulling	Carrying
....	....
....	....
....	....
....	....
....	....

Animals that pull are usually called 'draught' or 'traction' animals. Those which carry various kinds of loads are called 'pack' animals. Some animals do both types of work at different times. As a community facilitator, it will be important for you to find out which types of work are commonly done by working animals in your area, and the local names for the parts of their harnesses, saddles, packs, carts and other equipment.

How do people depend on working animals?

Question 3

Look around you again and think about your answers for Question 2. Describe how the work carried out by draught or pack animal contributes to people's lives. There may be several ways.

....

....

....

....

Working animals can help to earn money by transporting people, farm produce, animal feed, construction materials and many other things. They can also benefit families and communities indirectly, for example by relieving the burden of human labour, acting as a cheap alternative to expensive motorized transport or machines, providing social status and even boosting self esteem and happiness.

Question 4

Where people use working animals, what would happen to a family if their working animal was in a poor state and could not work? Would their lives be different if their animal was sick or if they no longer had a working animal? Write down how each family member's life might change. Ask animal owners nearby if they have experienced this situation and how it affected them and their families.

Husband..

....

Wife..

....

Girl child...

....

Boy child...

....

Elderly person...

....

Without a productive working animal, the whole family may have to work harder and longer to fetch and carry water, food, animal fodder, firewood and other goods. What answers were given by the people you talked to? Did their family members have less time to spend on other duties or on resting? Did the family have a lower income? Did they spend more money from the household budget on paying veterinary fees or taking a loan for a new animal? Were they less able to afford important things like children's education, or did their children have to spend their time on household work instead of going to school? Many people rely on their working animals for daily tasks or to help them earn an income.

How do working animals depend on people?

Question 5

Think about the working animals that you have seen. How can they meet their own needs for food and water, shelter and comfort, rest and relaxation, good health and the companionship of other animals? Do they need the help of people to get some of these things? Describe which ones and why.

....

....

....

....

Wild or non-domesticated horses, donkeys, cattle and camels carry out natural behaviour to look after themselves. They graze on plants in family groups, travel long distances in search of water and shade, and roll or rub against trees to remove dirt and parasites from their skin. Draught and pack animals work hard for people. This means that they do not have the freedom and time to carry out much of their natural behaviour. Working animals depend on people to get what they need and help them stay in a good, healthy state with enough energy to work productively.

Question 6

Look again at the animals around where you are sitting or walking. Think about the animals' working time and their resting time. Describe who working animals depend upon to stay in a good, healthy state with enough energy to work and be productive for the family. Who gives them feed and water? Who cleans the area where they are kept? Ask someone with an animal about what care it needs and who does it.

....

....

....

....

People have control over the lives of working animals. People choose when their animals can eat, drink and sleep, what they can do, where they can go, and which other animals they can meet. People are responsible for their animals' well-being because they use animals for their own benefit and this prevents the animals from looking after themselves.

How are people's lives influenced by the way that they care for their working animals?

Question 7

Watch the owners and users of working animals around you and think about other working animals you have seen. Take some time to consider how people think, react and behave towards working animals and how this can affect the animals' productivity and health. Then fill in the brackets with a plus sign (+) or minus sign (-) to show how the statement after the equals sign (=) was achieved.

- working animal ( ) responsible owner = more healthy and productive animal

- working animal ( ) available water = more dehydrated and unproductive animal

- working animal ( ) clean, dry saddle and harness padding = more comfortable and productive animal

- working animal ( ) veterinary check-up = more healthy and productive animal

- working animal ( ) beating = more painful, sad and unproductive animal

- working animal ( ) grooming = more clean animal which is free from disease and productive

- working animal ( ) shouting = more confused and anxious animal which is less productive

- working animal ( ) nutritious food = more hungry and unproductive animal

- working animal ( ) ability to move around = animal which is more comfortable and can perform some natural behaviours

- working animal ( ) wounds = more painful and unproductive animal

- working animal ( ) companionship of other animals = more comforted and relaxed animal

- working animal ( ) fear = more nervous, unhappy and unproductive animal

- working animal ( ) overgrown, dirty feet = more unhealthy, lame and unproductive animal

- working animal ( ) quiet and comfortable lying area = more rested, relaxed and productive animal

People's reactions and behaviour towards their working animals can have positive or negative effects on the animals' well-being and this affects their productivity. As you continue with the exercises, note down how people's actions influence their animals' health and ability to work.

Question 8

In this exercise, read the first half of the statement then fill in the second half, showing the benefits for (a) animal-owning families, and (b) the working animals themselves.

If working animals were healthier and lived longer:

(a) How would their owners and families benefit? ....

....

....

(b) How would the animals benefit? ....

....

....

If working animals were more productive:

(a) How would their owners and families benefit? ....

....

....

(b) How would the animals benefit? ....

....

....

If working animals were easier to handle:

    (a) How would their owners and families benefit? ....

    ....

    ....

    (b) How would the animals benefit? ....

    ....

    ....

If working animals recovered from stress and tiredness more quickly:

    (a) How would their owners and families benefit? ....

    ....

    ....

    (b) How would the animals benefit?...

    ....

    ....

If working animals had fewer injuries and days when they could not work:

    (a) How would their owners and families benefit? ....

    ....

    ....

    (b) How would the animals benefit?...

    ....

    ....

If people cared more about working animals:

    (a) How would their owners and families benefit? ....

    ....

    ....

    (b) How would their animals benefit?...

    ....

    ....

Every harm that people do to working animals, or positive action that they fail to take, can result in a problem for the animal and a loss for the owner and his or her family. An investment in caring for working animals and looking after them better will result in benefits for people and for the family's livelihood, as well as for the animal.

You should now be able to:

- state how working animals use their energy and power productively to help people;

- explain how wild animals can usually look after themselves, and why working animals are not able to provide themselves with everything they need;

- describe why being used for work prevents working animals from performing the natural behaviours that they would choose, such as grazing, finding water, playing and resting;

- discuss how working animals rely on people to consider their feelings and provide well for their needs;

- give examples of how working animals that are well cared for will be productive and contribute to the lives of the family and the life of the community.

Look after your working animals and they will look after you.

CHAPTER 2
Animal welfare

What you will find in this chapter

This chapter contains an introduction to animal welfare, including:

- what good welfare is;

- how to recognize it by observing animals and their behaviour;

- how welfare is affected by the different factors influencing animals' lives.

The chapter is organized in the same way as Chapter 1: first a question or short exercise to stimulate your thinking and draw from your existing experience, then a picture and a statement which answers the question and helps to test your new knowledge. The principles of animal welfare apply to all kinds of animals, so think about these questions and exercises when you are out in the street or the field and see animals of any species – working animals, farm livestock, pets or wildlife. This will help to expand your experience and generate new questions.

What is animal welfare?

Animal welfare is the term used to explain what animals need and how they feel. It is sometimes called animal well-being, state or quality of life. Animal welfare can be thought of as a continuum: it can be good or bad or somewhere in between. If animals are fit, feel good and have what they need, they have good welfare. The aim of this manual is to stimulate collective action to improve the welfare of working animals, taking them from whatever state they are currently in towards a better state.

After reading this chapter you will be able to:

- state why observation of animals is important in order to assess what animals need and how they feel;
- observe how animals interact with their environment, resources and people;
- list what 'inputs' are needed in order for working animals to be in a good welfare state at home, during work and during rest periods;
- give examples of how the needs of animals may differ or change according to their living, working and resting environments.

Look out for the theory boxes which will show you more ways to think about animal welfare.

Practise observing working animals

Watching animals closely (observation) is a great way to learn about them. For the next few exercises you will need to spend about one hour in an area where working animals are present. Walk around with this manual or your notebook, looking carefully at the animals and their surroundings.

Question 1

Observe an individual working animal or a group of animals for five to ten minutes. What did you notice about what was happening to the animals and around the animals? Were the things that happened good or bad for them? Were they things that the animals might like or dislike? Write down what you observed about the animals' behaviour and their reactions, including their interaction with people, resources and the environment.

What is happening? ....
....

Do you think it is good or bad for the animal? ....
....

What is the animal doing? Describe its behaviour and interactions..
....
....

What else is happening? ....
....

Do you think it is good or bad for the animal? ....
....

What is the animal doing? Describe its behaviour and interactions..
....
....

Observation can tell you many things about animals, including what they are doing and how they are doing it (animal behaviour). You can also observe how animals are affected by people's behaviour, resources and other things in the environment. Observing animals closely is important because it helps you to assess what animals need and how they feel – and to decide whether their welfare is good or if it could be improved.

What do working animals need?

Question 2

What do working animals need in order to be fit, healthy and productive for the families who own them? List all the things that you can think of.

....

....

....

....

....

Like people, animals need a variety of things to make sure that they stay healthy, feeling good and working productively. In order to improve animal welfare for the benefit of animals and people, it is important to know what working animals need in order to be fit and to feel good. The theory box on page 24 tells you what animals need for good welfare.

Theory box 3. The welfare needs of working animals

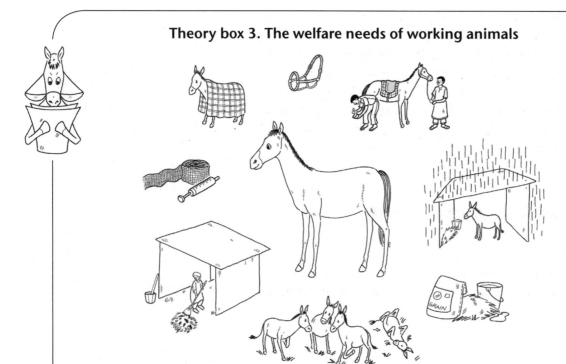

Draught and pack animals use up a lot of energy during a working day. They need adequate access to water and nutritionally balanced food each day, and sufficient time to eat and drink regularly. They need to feel relaxed in order to eat and drink enough to meet their needs. Good food and water give strength and energy, which ensure that animals can stay productive for their owners.

There is a limit to how much work an animal can do each day. If they are forced to do more work than they can mentally and physically manage, they will soon be in a poor welfare state. Enough good rest is another way to ensure that working animals can stay productive.

Working animals are on their feet all day, even when their owners are sitting or resting. If their feet are not looked after well, they will become damaged and painful. Regular foot care will help to keep animals working well for their owners.

Every type of work requires safe equipment to prevent animals from suffering pain and injury; for example carts, harnesses, ploughs and saddles. Equipment should be made with appropriate local materials, be well-fitted to the individual animal and always kept clean and dry.

Good health is an important part of good welfare. Disease influences welfare not only by its effect on the animal's body but also on the animal's mental state and feelings. Good health helps to create mental and physical stability so that the animal can function well in a variety of environmental conditions. Accessible, affordable and available health care and first aid are factors which help an animal to stay free from pain and distress and remain productive.

Good animal husbandry and management involves the provision of resources such as food, water and a clean, dry place to rest. It is also important to provide animals with appropriate handling and humane care during work and rest periods. Both resource provision and calm handling are essential to achieve a lifetime of good welfare. Animals are more likely to work effectively if they are well cared for.

Working animals need to conserve their energy in order to work well each day. If exposed to bad weather they will use up their energy trying to keep warm or cool, making them more prone to exhaustion and disease. They need a well-ventilated, sheltered resting area and protection from rain, sun, wind and extreme temperatures. A comfortable environment will keep animals healthy and working well.

Working animals are often deprived of their natural environment. This includes opportunities for self-care behaviours (such as rolling, running or scratching) and social interaction with other animals of the same kind (such as calling to each other or playing). Animals like to be near other animals and have natural curiosity. They enjoy the opportunity to inspect, manipulate, evaluate and interact with things in their environment and will choose to do this whenever they have the opportunity. It is important to give them an appropriate **physical and social environment** in order that they can feel safe and relaxed, which also helps them to stay productive.

Question 3

Watch some more animals for five to ten minutes. Which things that they need are they getting from their environment or the resources and people around them and which ones are they not getting?

Getting   Not getting

....

....

....

....

....

....

Although people and animals are not the same, they are living beings so have very similar basic needs. What you will need in order to feel good or work productively, a working animal is likely to need too. It is important to try to see the world through the animal's eyes. Think: 'If I was this animal, what would I want or need in this situation to ensure that I was healthy and feeling good?'

Question 4

How can we ensure that animals are getting what they need? Look at the pictures below and answer the question that goes with each one.

Why might it be difficult for this donkey to eat?

....

....

....

....

....

....

Why might it be difficult for this animal to rest?

....

....

....

....

....

....

What are some consequences for this horse if it doesn't have enough space?

....
....
....
....
....

What might happen to these animals if they don't have enough shade?

....
....
....
....
....
....
....
....

How might these bullocks be exposed to the spread of disease?

....
....
....
....
....
....
....
....

How might this donkey become injured?

....

....

....

....

....

....

....

....

....

Sometimes people make provisions for their animals, but the animals still cannot meet their needs. For example, people may provide food and water but tie up the animals so they are unable to reach it. Or one animal may have a sore mouth so that even when food is right there, it is unable to eat properly. In order to look after working animals and ensure that they can be productive for their owners, it is important to check that they are benefiting from the resources provided. This is why we try to see things from the animal's point of view. Not only do we ask: 'Is the owner providing inputs or resources?', but also: 'Is the animal really getting what it needs?'.

Question 5

Look again at the animals around you. Can you see any examples where a resource is being provided near to the animal, but the animal is not actually getting what it needs? Write down the reasons for this.

....

....

....

....

....

How do animals feel?

After completing the next set of questions you will know what animals can sense and feel, and also why animals' feelings are important for their welfare.

   People and animals are similar in their needs and also in their basic feelings or emotions. They are likely to feel quite like you about things which affect their needs and their daily life. Keep practising trying to see the world through the animal's eyes. Ask yourself: 'If I was this animal, how would I feel about what is happening here? How would I feel about getting this need or not getting it?'

Question 6

Watch another animal for five minutes, preferably one which is with a person. First write down how you think that animal feels about getting or not getting its needs from the surrounding environment.

Needs from environment	Getting them or not?	How might the animal feel?
....	....	....
....	....	....
....	....	....

If you were the animal, what (if anything) would you want to avoid in the environment or surrounding area? Why?

....

....

Now consider how the animal feels about its interaction with the person it is with. Think: 'If I was the animal, how would I feel about this person? How would I feel about what they are doing to me, or for me?'

Needs from person?	Getting them or not?	How might the animal feel?
....	....	....
....	....	....
....	....	....

If you were the animal, what (if anything) would you want the person to stop doing? Why?

....

....

One way to decide whether a person's action towards an animal would make it feel good or bad is to observe **how** the action is being done. Choose your own words which best describe what is happening in the interaction between the person and the animal. For example: gently, firmly, quickly, slowly, unkindly, kindly, hurriedly, maliciously, impatiently, angrily or considerately. This helps you to think of **the way** in which something happens to an animal and whether this would make the animal feel good or bad.

The next step is to think: **why** is the animal getting or not getting its needs? Why is the environment positive or negative for animal welfare? Why is the person acting or not acting in a particular way towards their animal? Why are they providing or not providing a resource? There can be many reasons!

Question 7

Look at your earlier observations. List some reasons why these things happened to the animal. You may want to ask the person who is with the animal, to see if the reasons they give are the same as yours.

....

....

....

....

....

Question 8

Now that you have completed some animal observations, have you thought of any more things that animals need in order to be fit and feel good, which you did not consider before?

....

....

....

....

Question 9

Go outside, walk around and have a look at some more working animals in the village, market or street. Observe one of the animals in particular. If it was a wild animal and could choose anything that it wanted to do, what do you think it would naturally choose? Then think about whether, as a working animal, it is being treated in a way that it would choose in the situation that it is in right now. Is it likely to be feeling something that it wants to avoid, such as pain, fear, thirst or heat? Does the animal want something that it doesn't have? Is it deprived of a need, such as food, companionship or the ability to stretch, scratch or rest?

....

....

....

....

....

Wild animals have a lot of choices about how they spend their time. Even farm animals can often choose where and when to walk, scratch, eat or lie down. Working animals often have a very restricted and unnatural life. The cart, plough or pack load deprives them of their freedom to interact socially with other animals, explore their environment and behave in the way they would choose to do. The long hours of work put a lot of stress on them; they have little rest and may be overloaded, beaten or roughly handled. People are responsible for the life experiences of working animals.

It is possible for people to help animals to avoid the things they dislike and offer them opportunities to do what they would naturally choose to do.

Question 10

Animals constantly gather information from all their senses (sight, smell, taste, hearing and touch) to inform them about their surroundings. Look at several animals nearby. Who or what in their surroundings are they gathering information about? How are they gathering the information? How can you tell?

....

....

....

Why might the information be important to them?

....

....

....

....

....

Wild animals use information from their senses to help them find food, water and shelter and to avoid predators. Their first priority is to survive, so they look for changes in their environment and determine what is safe and what is threatening. Domesticated working animals use the same senses to help them stay safe, to live comfortably with other animals and people, and to look after their own needs as much as possible.

Theory box 4. Prey species and animal senses

Working animals are all prey species – in the wild they would be hunted and eaten by predator species such as lions and wild dogs. In order to survive and avoid predators, prey species normally live in social groups and have very well-developed senses which inform them about their surroundings. Their senses tell them:

1. What is present in the environment?
2. How much is there?
3. Where is it? ('It' can be anything present in the environment).
4. Is it moving or changing?
5. Is there more or less of it than before?
6. Is it good or bad for them?

Animal senses:

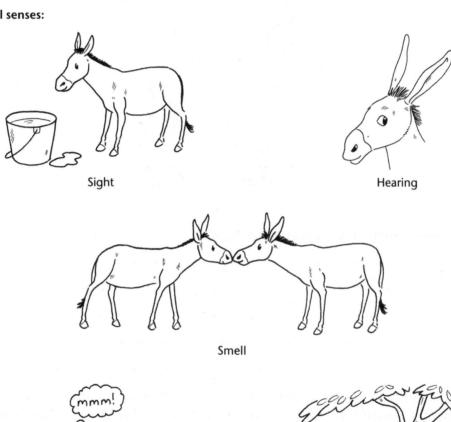

Sight

Hearing

Smell

Taste

Touch

When prey species sense something in their environment which is threatening, they are frightened and want to run away with their group. If working animals cannot run away and do not have a social group to make them feel safer, they may stand very still and stiffly, or may try to defend themselves by kicking or biting. This is normal fear behaviour. By being aware of their natural behaviour, you can find ways to prevent or reduce their fear and help them to feel safe.

Question 11

Spend ten minutes looking again at the working animals near you. How do you know that the information gathered through their senses leads to them having feelings about people, other animals and the environment? List as many clues as you can find that the animals may have feelings, based on what is happening to them or around them. What feelings are they showing and how are they showing them?

....

....

....

....

....

....

When animals sense or experience things in their surroundings, this generates feelings about their environment and the people or other animals in it. As with people, these feelings can include pleasure, relaxation, fear, anxiety, frustration and many others. In the wild, these feelings cause animals to behave appropriately in order to meet their needs, protect themselves and maximize their chances of survival.

Domesticated animals (including working animals) also have these feelings about their environment, people and other animals. If they can behave appropriately to their feelings, such as eating the food they smell, running away from dangerous sounds or playing and socialising with each other, they will feel good and so their welfare will be good. However, working animals can be vulnerable if their living or working situation does not allow them to react to their feelings. For example they may smell danger but be unable to run away or seek the safety of companions. Or they may see food and be hungry but be unable to reach it. The environment of a working animal can help it to feel good, or can present a challenge, depending upon how the animal is allowed to interact with its surroundings. When working animals are unable to behave or respond appropriately this leads to poor mental or physical welfare: they feel bad.

How can you tell if animals feel good and have what they need?

At this stage you may be wondering how to tell whether a working animal's needs have been met, so that it feels good and can maintain a productive working life. In Question 6 you considered how you would feel if you were the animal, but how does the animal tell you how it really feels?

Animal behaviour is the expression of animal needs and feelings. At the end of this section you will be able to describe what an animal's behaviour can tell you about its feelings and welfare state.

Question 12

How might a working animal behave or react in response to the following situations? What feelings might the animal experience that would cause it to react in the way you described?

(a) Noise?

Reaction/behaviour...

Feeling that leads to the reaction...

(b) Being alone, or isolated from other animals of its own kind?

Reaction/behaviour...

Feeling..

(c) Seeing familiar people or animals?

Reaction/behaviour...

Feeling..

(d) Seeing unfamiliar people or animals?

Reaction/behaviour...

Feeling..

(e) Being handled in a calm, gentle way?

Reaction/behaviour..

Feeling that leads to the reaction..

(f) Being handled in a jerky, angry way?

Reaction/behaviour..

Feeling that leads to the reaction..

(g) Something frightening coming closer to it

Reaction/behaviour..

Feeling that leads to the reaction..

(h) Something frightening staying far away from it?

Reaction/behaviour..

Feeling that leads to the reaction..

(i) Travelling to a new location?

Reaction/behaviour..

Feeling that leads to the reaction..

(j) Doing a new type of work?

Reaction/behaviour..

Feeling that leads to the reaction..

(k) Having cool water to drink on a hot day?

Reaction/behaviour..

Feeling that leads to the reaction..

Behaviour is the way in which an animal expresses its feelings. If it feels frightened, it may show this through behaviour such as putting its ears back, standing very still, kicking or trying to run away. This is sometimes called 'body language'. It is important that you can understand from their behaviour what animals are trying to tell you about their feelings. If you recognize that they are feeling bad, you can start to improve their welfare by meeting their needs. When they start to feel better, you will also see this expressed in their behaviour – they will look relaxed with their ears forward and show calm behaviour towards people and other animals.

Question 13

Now that you have more information and have started to gain experience of looking at working animals, go outside and observe them for another five to ten minutes. For each animal you see, first write down what the animal is doing, for example, pulling a cart. Then observe how or in what way the animal is doing it. For example, in what way is the donkey carrying the goods or pulling the cart? Add describing words, such as pulling easily, stiffly, comfortably, struggling, stumbling.

What is the animal doing?

....

....

....

....

....

....

How is it doing it?

....

....

....

....

....

....

Theory box 5. Listening to the 'voice' of the animal – unwanted behaviour

Working animals can sometimes behave in a way that their owners and users do not like, such as biting, kicking, running away or refusing to move forward. People may say that the animal is being stubborn, aggressive or vicious. Remember that these behaviours are normal – they are the **animal's way** to show how it feels, protect itself and feel safe. They indicate that the animal is not fit or not happy.

It is important to listen to the 'voice of the animal' when it behaves in this way. Encourage the owner to check for injury, illness or disease. Check for discomfort or pain from its harness, pack or other equipment. Check for fear or distress as a result of other animals around it, the environment, a resource issue or a person's behaviour towards the animal. If none of these are present, it may be that the animal does not understand what it is being asked to do. With **kind training, good management and calm handling**, these behaviours will disappear, making life easier for both the animal and the people who work with it.

If you practise looking at **the way** in which animals behave when they experience things in their lives, you will begin to be more aware of the animal's point of view. You will become more sensitive to small changes in the body language and even the facial expressions of animals. Eventually you will find that it comes naturally to interpret what an animal's behaviour tells you about its feelings, and to see the world from the animal's point of view.

What are the signs of good and poor welfare?

You can tell how working animals **feel**, physically and mentally, by looking carefully at physical signs on their bodies and observing their behaviour. After the next set of questions you will be able to identify physical and behavioural signs of good and poor welfare.

Question 14

If an owner understands and provides what his or her animal needs, what signs would you see on the animal's body which indicate good **physical** welfare?

....

....

....

If its owner does not provide what an animal needs, what would you see on its body that indicates poor physical welfare?

....

....

....

....

Look around you. What signs of good or poor physical welfare can you see on animals in the market or street? Write them down here.

....

....

....

....

An animal that is not fit and healthy will inevitably be suffering from poor welfare. To be fit, the animal needs to have good health and vigour throughout its working life. If welfare problems are prevented and the needs of working animals are met, they are more likely to show positive physical signs, such as good body weight, clean coat and feet, strong movement, energy to interact with the surroundings and willingness to work. They are more likely to feel good. As you saw in Chapter 1, working animals are dependent on people to provide their care. When people no longer provide for their needs or treat them well, working animals become unfit and unhealthy.

Question 15

If its owner understands and provides what a working animal needs, what signs from the animal would indicate good **mental or emotional** welfare?

....

....

....

....

If its owner does not provide what the animal needs, what signs would indicate poor **mental or emotional** welfare?

....

....

....

....

Did your answers include examples of behaviour or body language? Remember that behaviour is the way in which the animal expresses how it feels – it is the 'voice' of the animal. Look at the animals nearby again. What signs of good or poor **mental and emotional** welfare can you see? Write them down here.

....

....

....

....

An animal which is not mentally calm and happy will also be suffering from poor welfare, even if its body is healthy. Indicators of positive mental and emotional welfare include: alertness, a relaxed posture and facial expression, playing, investigating surroundings and interacting in a friendly way with people and other animals of the same kind.

Like people, animals need to be in both a good physical state and a good mental (or emotional) state in order to feel good. Having either one or the other is not enough for good welfare. To feel good, working animals need to be in good physical condition and have fewer episodes of disease or injury. They also need to experience plenty of positive emotions, and live through fewer negative emotional experiences. If people care about their working animals and provide for their needs (inputs), working animals are more likely to have good feelings such as pleasure, comfort, satisfaction and companionship (positive outputs) and to avoid bad feelings such as frustration, stress, pain, fear, discomfort and loneliness (negative outputs).

Question 16

Animal needs can be thought of as the 'inputs' into the animal – the resources and management practices which help it to achieve good welfare. The signs and indicators of good welfare seen on the animal are the 'outputs'. Which of the animals in the following picture will support its owner's livelihood best? What was it about the animal's needs (inputs) and the signs or indicators seen on the animal (outputs) that led you to this decision?

....

....

....

....

....

....

....

Considering and providing for the animal's needs and feelings as much as possible will help to prevent welfare problems from developing. The most powerful way to improve animal welfare is to prevent problems before they start. Part of your role as an animal welfare facilitator will be helping people to recognize the early signs that an animal feels bad and take action quickly to improve welfare.

Question 17

How much or how many inputs or resources (food, water, rest and all the other animal needs) are necessary for a working animal to have good welfare?

· ·

· ·

· ·

It is not usually practical or realistic to give a working animal absolutely everything it needs to keep it feeling good all the time, every day, every season, throughout its life. However, the more of the animal's needs that are met and the better its feelings are understood, the better its welfare will be. Increasing good resources and management practices and decreasing poor resources and management practices will make a difference to the animal. Some changes will make an immediate, short-term difference, while others will make a sustainable, long-term difference. All of them are valuable and even small steps are better than no action at all.

Theory box 6. Animal welfare frameworks

Animal welfare can be a complex subject. Several animal welfare scientists have created frameworks for defining and understanding welfare in ways which are simple to understand. As there is no single 'right answer', it is useful to know some of the most common frameworks in order to have different perspectives on animal welfare for your work as a facilitator.

Fit and feeling good

This definition says that an animal has good welfare if it is 'fit and happy' or 'fit and feeling good'. Fitness means that the animal can sustain health and vigour throughout an effective working life. 'Feeling good' recognizes that animals are sentient, in other words they have feelings that matter. We should aim to ensure that they do not suffer and have the positive feelings gained from comfort, companionship and security.

Physical welfare, mental welfare and naturalness

This view of animal welfare emphasizes three components. Like 'fit and feeling good', it recognizes that both physical and mental welfare are important. It also includes 'naturalness': the ability of an animal to do what it would choose to do in a natural or wild state. For example, the opportunity for a working donkey to be still a donkey – grazing, braying, socialising in a herd with other donkeys – rather than just a machine for people's benefit.

Five freedoms

This framework looks at welfare outputs in terms of 'freedoms': ideal situations for animals which we should work towards achieving. Each freedom is then linked to the inputs (resources and management practices) which are needed to reach that freedom.

- *Freedom from hunger and thirst* through ready access to fresh water and a diet to maintain full health and vigour.
- *Freedom from discomfort* by providing an appropriate environment including shelter and a comfortable resting area.
- *Freedom from pain, injury or disease* through prevention or rapid diagnosis and treatment.
- *Freedom to express normal behaviour* by providing sufficient space, proper facilities and company of the animal's own kind.
- *Freedom from fear and distress* by ensuring conditions and treatment which avoid mental suffering.

After you have completed Chapters 1 and 2, see if you can identify how we have used each framework in the exercises.

Further reading

Webster, A.J.F, Main, D.C.J., Whay, H.R. (2004). Welfare assessment: indices from clinical observation. *Animal Welfare 13, S93-98*

Fraser, D., Weary, D.M., Pajor, E. A., Milligan, B.N. (1997). A scientific conception of animal welfare that reflects ethical concerns. *Animal Welfare 6, 187-205*

Five Freedoms. (1979). Farm Animal Welfare Council, UK.

How does welfare change over time and in different situations?

When you have finished the next three questions you will know how animal welfare changes in different situations and over short and long periods of time.

Question 18

Think about the working animals you have observed today and others you have seen or heard about in the past, living in different places and doing different kinds of work. List some environments where animals live and work. How might the needs of animals change in these environments? Would they have different or extra needs in order to stay healthy and productive for their owners?

Living or working environment	How this affects animal needs
....	....
....	....
....	....
....	....
....	....
....	....

The resources and management practices required in order to meet the needs of working animals may change depending on the situation that they are in at any particular time. During your work as a facilitator, it is important to become familiar with the variety of environments in which animals live, work and rest. You may wish to repeat the exercises above in several places and at different times, so that you can see these different needs for yourself.

| 0 years | 3-5 years | 5-8 years | 12 years |

Question 19

What are the changes and challenges to the welfare (needs and feelings) of working animals at different stages in their lives? Why do these changes happen? How do their experiences at work and at home affect them as they get older?

....	....	....	....
....	....	....	....
....	....	....	....
....	....	....	....
....	....	....	....
....	....	....	....

Just like people, animals experience mental and physical changes and challenges during their lifetime. The needs and feelings of animals are likely to change as they reach middle and older age. There are many reasons for this, such as changes in nutritional needs, working ability and the amount of rest or care needed by the animal and actually provided by its owner.

Question 20

Now consider the changes and challenges to working animals' welfare (needs and feelings) that you might expect to see in different seasons during one year. Why do these happen?

Ploughing Season   Harvesting Season   Hiring season

....

....

....

Within a single year, people and their working animals will experience many changes and challenges according to the climate, work load, food availability, income and other livelihood and environmental factors. In different countries and regions there will be different **seasons** or times of change within a livelihood system in one year and these will affect the welfare of working animals.

Question 21

Consider the changes and challenges to the welfare of working animals that can occur during a single working day. Why do these happen?

....

....

....

Within a single day, working animals will experience changes and challenges to their welfare. These will be influenced by their living, working and resting conditions and their health status. They will also depend on the opportunities to experience positive feelings that are provided by their environment and the people and other animals around them.

People understand how their own needs and feelings change as they get older, in different seasons of the year, and during a single day. It is a good rule of thumb that when the needs of animal owners change, the needs and feelings of their working animals will also change. Animal welfare is not a static condition, but a changing, dynamic situation. Your role as a facilitator is to enable animal-owning communities to commit to improving welfare over the course of each day, each season and the lifetime of the animal. This will ensure that their animals live longer and remain productive for the family.

Who knows about and influences the welfare of working animals every day?

After the last two questions you will have learned who knows most about working animal welfare on a day-to-day basis and why this is important to you as a facilitator of animal welfare improvement.

Question 22

What will happen to a working animal if its needs and feelings are not cared for? What will happen to its owner and his or her family?

....

....

....

Without good care, animals quickly become sick, weak, unhappy and unproductive. A working animal in a poor welfare state cannot thrive and provide a family with income in the way that a fit and healthy animal does.

Question 23

Who knows the most about a working animal's daily life and about its needs and feelings? Why?

....

....

....

....

Animal owners and their families are with their animals for long periods of time every day. They know when working and living conditions change for themselves and their animals. They know that when animals are in a poor welfare state they are less productive than when they are fit and happy. As a facilitator, you will not see the daily, seasonal and lifetime changes in animals' needs and feelings which occur within different livelihood systems. Veterinary and animal health workers, harness-makers, farriers and other stakeholders do not have such detailed knowledge either. Animal-owning families are the people who are most familiar with the indicators of physical and mental welfare shown by their own animals. They are the people who can make the biggest difference. With your facilitation, animal owners, users and carers can share their real daily experiences. Then they can use this wealth of experience to plan for collective action to improve the welfare of their working animals.

You should now be able to:

- observe working animals and describe their interactions with people, resources and the environment around them;

- describe the senses of animals and why animal feelings are important;

- list the resources and management practices that working animals need to help them feel good and stay productive;

- recognize signs of good and poor welfare by looking at animals' bodies and behaviour;

- describe how the welfare of working animals can change over time and in different situations;

- explain why it is important to work with animal owners, users and carers in order to improve welfare.

In Part II we will explore in detail how to facilitate groups of animal owners, their families and other stakeholders to improve the welfare of working animals. If you would like to read more about animal welfare before continuing on to Part III, have a look at the further reading and reference list at the back of this manual.

Case study A. Our donkeys live longer now!

Source: Mohamed Hammad, Ahmed El Sharkawy and Amro Hassan, Brooke Egypt, January 2010

In the Helwan region near Cairo, a huge collection of brick kilns produces 200 million red bricks every month. The kilns use over 1500 donkeys and 324 mules to pull brick carts. Brick kiln donkeys have many welfare problems, including dehydration, poor body condition, foot problems and wounds from saddles and from beating. Due to the harsh conditions, their mortality rate is high and many donkeys die young.

The Brooke's Community Mobile Clinic Team visit Helwan regularly to provide veterinary treatment and organize animal handling and husbandry training for animal caretakers and brick cart drivers, who work with the donkeys and mules on a daily basis. The team also facilitates meetings with the brick kiln owners, to discuss the benefits of looking after working animals well.

In 2003 an assessment of welfare was carried out on a random sample of animals in the brick kilns. Only 17 per cent of the donkeys were aged over 16 years. This welfare assessment was repeated in 2009 and the number of donkeys aged over 16 years had increased to almost 40per cent.

The community mobile team met with several factory and animal owners to find out if they had noticed this change and what they perceived to be the main reasons for it. Brick factory owners recognized that the turnover of working animals had sharply decreased. Now their animals live for longer and produce more income which makes them more valuable. Abdul Satar, a 45-year-old factory owner said 'The longevity of our donkeys has increased through provision of timely treatment and by keeping the stables clean, with water troughs inside'. Another factory owner mentioned that with the help of the mobile team he had improved feeding in the donkeys' peak working season, based on what was available and affordable in the area. He said 'Our donkeys are fed properly throughout the year, no wonder they live longer now!'

Emad Abo Ghorieb, the 32-year-old owner of several donkeys, explained that water was a problem. In the peak season his donkeys were suffering. He worked out how to provide more water for the donkeys by moving the water source inside the stable, giving more opportunities for the animals to drink. He said 'Right after that I noticed an improvement in my animals' health and felt so happy'.

Over the years these very small adjustments have made a real difference to the welfare of the donkeys, reducing mortality rates and prolonging their lives at the brick kilns.

PART II
IMPLEMENTING ANIMAL WELFARE INTERVENTIONS WITH COMMUNITIES

What you will find in this section

- Chapter 3 explains the types of animal welfare interventions that give lasting results. It describes how to identify the groups of working animals that are most in need of welfare improvement and how to decide which interventions are most appropriate for different groups of animals and owners.

- Chapter 4 contains a step-by-step guide for setting up and facilitating community groups to improve animal welfare.

- Chapter 5 describes methods for outreach and delivering welfare messages to scattered or less accessible populations.

How to use this section

The information in this section is important for planning your work with animal-owning communities. We suggest that you read Chapter 3 and decide which communities to work with and how. Then use Chapter 4 or Chapter 5, depending on which is most relevant to your decision. If you plan to work with several communities in different ways, both chapters will be useful.

CHAPTER 3
Interventions for lasting change

What you will find in this chapter

In this chapter we look at how welfare improvements can be made to last and can be supported by other stakeholders as well as the animal owner and his or her family. We also show you how to use your knowledge about animal welfare developed in Chapters 1 and 2 to plan how you will work with different communities and their animals.

What makes an intervention succeed?

In Chapter 2 we showed that the welfare of an animal is not fixed or static, it is a dynamic, changing situation. Facilitating owners to improve the welfare of working animals to an acceptable level at a certain moment in time does not necessarily mean that long term improvement in their welfare will be maintained. The welfare status of an animal changes daily, seasonally and over its lifetime, as a result of changes in its environment and in its living and working conditions. Any intervention aimed at sustained improvement needs the community to recognize when their animals' welfare is getting worse and to take action quickly.

Theory box 7. Incremental change

Animal welfare improvements are progressive and can be achieved by considering what working animals need and how they feel, then making small changes one after another. These are sometimes known as incremental changes or 'bite-sized chunks'. Welfare is not an all-or-nothing state. There is value to changing the lives of animals for the better, one step at a time.

As a facilitator, how can you help people to decide how much improvement they can make in the welfare of their animals, within the constraints of the resources and time available to them?

Encourage the group of owners to come to a collective agreement on how to define good management practices. Depending upon the local situation and the actual facilities available to them, they can decide what changes they will make and which changes to make first (the tools described in Part 3 will help with this). This usually means making relatively small improvements to the current situation, such as adding one or two good things to the animals' lives and reducing or taking away one or two things which make animals' lives harder.

For example, we have heard many owners say that is difficult to give sufficient space to their animals during rest periods or during the night, because there is not even enough space available for their own family members to rest. In this situation, we encourage the group to decide together on what small improvements can be made and maintained in the real life situation. When these succeed, the process of action and reflection that you are facilitating will generate further incremental steps towards practical, sustainable improvement in animal welfare.

Determinants of welfare

The welfare status of working animals varies depending on a complex range of factors influencing its life. These factors can be called determinants and are presented in the figure below.

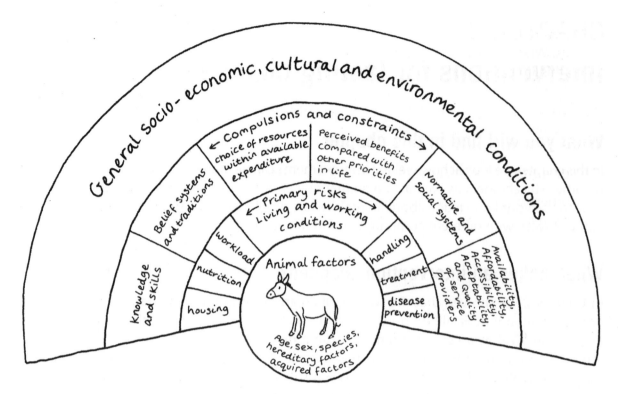

Figure 3.1 Determinants of working animal welfare. Source: van Dijk, L. and Pritchard, J.C., (2010), adapted from Dahlgren and Whitehead, (1991)

At the core of the diagram is the animal itself. Its welfare is partly influenced by 'animal factors' such as its age, sex, species and the features inherited from its parents (for example a particular body shape) or gained during its life up to now (such as fear of people).

Moving outwards, the second layer of the diagram shows the external factors which affect the welfare of the animal directly, on a daily basis. These are its living and working conditions, such as housing, nutrition, workload, handling, disease prevention and treatment.

The third layer shows the factors which influence the animal's living and working conditions and therefore determine its welfare indirectly. These are called compulsions and constraints. They include the knowledge and capacity of the people who deal with the animal, the services available (including animal health services), the resources available in the locality and, among those, the specific resources that owners choose to provide, depending on what they can afford. This layer also contains the belief systems and traditions of the people dealing with the animal, the influence of their peers and social network, their social status and their income level.

The outer layer represents more general socio-economic and environmental factors which have an effect on the third layer. These may include social structures, droughts and floods, mobility patterns, urbanisation, fuel prices and changes in mechanised transport.

Within this diagram, the factors or determinants of animal welfare influence each other within each layer and between one layer and the next. In order to be successful in improving welfare you will need to address several factors at the same time: the animal, the people dealing with the animal and the systems in which they both live and work. Some factors influencing animal welfare are within the owner's control, such as whether it is beaten, or what time of day it is fed. However, many factors cannot be influenced by individual animal owners because they are part of a wider living and working system, or socio-economic system. Only a collective approach by many people at the same time will be able to solve these issues.

For example, in many countries the social status of owners can influence the welfare of their animals. Where people are marginalised from society - such as lower-caste donkey owners in India - this limits their ability to care for their animals because service providers and government officials will not listen to the problems of a single person or family. Unless these owners are able to work collectively to influence others, they are not likely to see any long term welfare improvement in their working donkeys.

Cornerstones of a successful, lasting welfare intervention

The cornerstones of a sustainable intervention for improving the welfare of working animals are shown below:

Motivated and knowledgeable animal owners, users, carers

Quality local service providers and resources available

Interventions for lasting improvement in the welfare of working animals

Strong, cohesive group structures

Mechanisms for monitoring welfare in the community

Figure 3.2 Cornerstones of a successful, lasting welfare intervention

Process box 1. Sustainability mapping

The illustration of the cornerstones of a successful lasting welfare intervention (Figure 3.2 above) is an example of a sustainability map. It enables people to visualise the systems, mechanisms, institutions and processes (sometimes called 'results areas') that need to be in place in order to achieve a long-lasting improvement in working animal welfare.

- Community members can develop a sustainability map to identify what they need to do in order to maintain good working animal welfare on their own, without the support of external agencies. This can be used to prepare the community for gradual withdrawal of external support from a community (see Chapter 4, Phase 6).

- Our field managers and community facilitators also find sustainability mapping to be effective for planning their own projects. We have used it during annual planning workshops to explore the specific community engagement situations needed to achieve our objectives of facilitating improved welfare for working animals.

After initial identification of these systems, mechanisms, institutions and processes, the exercise then breaks down each results area into the specific activities needed to achieve sustainability.

Step 1	Ask the group what systems, mechanisms, institutions and processes would need to be in place so that their current animal welfare activities could continue on their own without external facilitation or support. Give the participants coloured cards and ask them to draw or write down their thoughts.
Step 2	Enable them to sort out all these cards into categories and paste them onto a big piece of chart paper. Build consensus within the group by facilitating thorough analysis and debate about what is shown on each card. Based on all participants' contributions, develop common statements about each results area.
Step 3	When results areas are finalised, encourage participants to identify the activities which would be needed to achieve each result area. Ask them to draw or write each activity on the chart under the result area which it contributes to, and discuss the opportunities for carrying out these activities. Enable the group to develop an action plan based on the activities identified.

Sustainability mapping is slightly different from the Vision or Dream mapping that you may have seen in other contexts (Kumar, 2002). In Vision mapping, broad goals are visualised and vision statements are often drawn by participants in the form of pictures. Sustainability mapping is more useful for enabling participants to identify specific results areas and activities and to form a concrete action plan.

Motivated and knowledgeable owners, users and carers

There are several ways in which you can motivate working animal owners, including:

- *Reinforcement of the value of healthier animals*. This can be done by discussing the economic contribution of working animals to their owners' household income, or by comparing the relative longevity, ability to work and cost of veterinary care for animals with good and poor health. To discuss these in detail, use the How to increase the value of my animal tool with owners (see Toolkit 18).

- *Choosing the welfare intervention most preferred by the community* as a starting point, based on what they feel is most important for themselves and their animals. Creating opportunities for animal owners, users and carers to identify and act upon the most pressing issues affecting the welfare of their animals can help to build their active involvement and ownership of the intervention.

- *Peer pressure from other owners* is very useful in creating motivation. This is best achieved through formation of a group for collective action (see below).

- *Competitions*. We have found village animal welfare competitions such as happy donkey competitions, to be very motivating in our work in India, Pakistan and Kenya. In these, animal owners jointly set the criteria for winning animals and judge the competition together. See Chapter 5 for more details and a case study on village-to-village competitions.

- *Use of existing local knowledge and practice*. Many welfare problems identified by the owners can be solved by unlocking and sharing their existing knowledge through discussion. Women usually play a significant role in the care of the working animals at home. They have their own knowledge about animal issues which the men may not realize. For lasting change, women need to be involved in welfare improvement interventions in the most appropriate way, whenever they are involved in the care of animals.

If you are concerned that some existing practices may be causing harm to the animals, discuss this with your fellow facilitators or a local animal health provider. You can then use tools such as Animal welfare practice gap analysis (see Toolkit 21) to explore these problems with the community.

Formation of a strong, cohesive group

Forming a strong and cohesive group is essential to achieving the motivation, knowledge and monitoring mechanisms for mutual learning and peer pressure to improve the lives of working animals. As described in the Introduction to this manual, each group builds their capacity to identify welfare issues and act on them. Co-operation between owners allows the group to do things that its members could not achieve as individuals, such as buying animal feed in bulk and holding service-providers to account for the quality of service they provide. Any form of collective structure to improve animal welfare is likely to be more sustainable than individuals trying to act alone . See Chapter 4, Phase 1: Feeling the pulse, for detailed information on formation of groups.

Mechanisms for monitoring animal welfare status by the community

Improving animal welfare and sustaining this improvement requires people to prevent welfare problems, to recognize any deterioration in welfare at an early stage, and to have the capacity to act in order to reduce the negative impact. If all of these are in place, welfare problems affecting working animals will become less frequent and less extreme. Developing a system where the owners themselves monitor the welfare status of their working animals together on a regular basis is essential in order to make a lasting difference, leading owners to make changes quickly for their animals' benefit. Motivation is also strengthened by peer pressure: the knowledge that others in the group will see the changes made and monitor the welfare of all animals belonging to the group. We describe these mechanisms further in Chapter 4.

Quality local service providers and resources are available

Although animal owners and their families make the biggest long-term difference their animal's lives, other stakeholders or service-providers also play an important role in the welfare of working animals.

As a facilitator, you can help owners to identify the most suitable resources for their animals and help to link them with important service providers who will attend to their animals' needs, such as cart-makers, animal health workers, feed-sellers and medicine shops (see Figure 3.3 below). In some places these service providers are members of the animal-owning community and they may be part of the group, or may be invited to meetings and kept informed about the activities and progress of the group by the members.

TYPES OF SERVICE PROVIDER	EXAMPLES OF HOW THEY MEET THE NEEDS OF WORKING ANIMALS
Saddler and harness-maker	makes, sells and repairs saddles and harness
Feed-Seller	Sources and sells animal feeds, chops and mixes different types of feeds
Farrier or blacksmith	Trims feet, makes and fits shoes
Vet, paravet or animal health technician	Diagnoses and treats sick animals
Community animal health worker	Diagnoses and treats sick animals vaccinates animals, reports disease outbreaks to local authorities
Agrovet or medicine store	Sells animal medicines and equipment advises on what medicine to use

Figure 3.3 Types of service provider and how each one helps to meet the needs of working animals

Theory box 8. AAAAQ of service providers

Local service-providers play an indispensable role in enabling the community to care for their animals. In almost all communities, local service-providers are already available and providing services to working animals to a greater or lesser extent. Examples include farriers (blacksmiths), feed-sellers, cart- and harness-makers and repairers, and health service providers such as government or private vets, Community-based Animal Health Workers or local traditional healers.

There are two fundamental driving forces behind effective and lasting service provision:
- **Demand** The animal owners perceive a need for the services and are therefore willing to seek them out and pay for them.
- **Supply** A service-provider can make a living from providing a service to the people who need and want one.

To be successful, any services should meet the following criteria:

Accessibility The service should be accessible to the local community- this is a key component of sustainability (including an accessible medicine and equipment supply in the case of health service providers). Emergency treatment of working animals requires a fast response.

Availability The service provider needs to be available when required. Flexible working hours are important, such as availability at night and when owners are not working themselves.

Affordability If the community cannot pay, they will not use the service and the service-provider will not be able to continue to provide it. This often includes the flexibility to pay in kind and to get some services on credit. Affordability also depends on how animal owners judge the cost of the service in relation to the skills of the provider.

Acceptability The service-provider is more likely to be accepted by the community if they are actively involved in his or her selection. Examples of acceptance indicators are confidence and trust.

Quality Quality is important in terms of provision of an appropriate, welfare-friendly service to animals. If the community does not value the quality of the service, they will not use it.

Adapted from: Catley, A., Blakeway, S. and Leyland, T. (2002). *Community-based Animal Healthcare: A Practical Guide to Improving Primary Veterinary Services*. BookPower/ITDG Publishing, Rugby, UK.

Case study B. Involvement of key stakeholders in improving animal welfare

Source: Ratnesh Rao, Brooke India, Bijnor, Uttar Pradesh, India, September 2009

Rawati is a village in the Bijnor district of Uttar Pradesh, which has 21 households with 40 horses and donkeys. The animals are engaged in work for brick kilns and potteries during the summer and winter seasons and for the rest of the year they transport other goods and people. Initial contacts by the Brooke India's Bijnor Equine Welfare Unit laid the foundation for an Equine Welfare Group which was formed in July 2008. The group started a monthly collective savings scheme. Loans taken from this have been used for purchase of feed, paying for treatment, buying horses and donkeys, repairing carts and domestic household purposes.

The group recognized that animal-related stakeholders – such as farriers, hair-clippers, feed-sellers, veterinary service-providers and medicine shopkeepers – play an important role in the welfare of their animals. Tools such as Cost-benefit analysis (Toolkit 15) and Pair-wise ranking (Toolkit 8) helped them to analyse the constraints and opportunities relating to their use of the stakeholders' services.

Hashim, the local farrier, was invited to join the group when it was formed. This provided the opportunity for in-depth analysis to improve the quality of horseshoes and reduce the costs of shoeing. The group decided to purchase good shoes from the local market using their savings fund, and also negotiated with Hashim to reduce his charges for members of the group. Hashim charges 60 Indian rupees per horse for a group member and 80 rupees for the same service outside the village. This high quality service at reduced cost has led to more regular shoeing of the animals and fewer hoof problems. Although the price he charges per horse is lower than before, this arrangement benefits Hashim, because people bring their animals for shoeing more regularly and have all four feet shod together instead of one or two at a time.

The hair-clipper was only available far away from the village and a lot of constraints were identified with use of this service to clip the coats and manes of the animals during hot weather. Collective analysis of the problem identified various alternatives. Three group members came forward to start clipping hair, not only for animals belonging to the group but also in other areas as an income-generating activity. They purchased hair-clipping machines by taking loans of 2000 rupees each from the savings fund, charging 50 rupees to clip a group member's horse and 70 rupees for everyone else.

The Equine Welfare Group has also established links with local veterinary service-providers (Raghunath and Suresh), medical stores (Akshay and Vipin Medical Stores) and a cart maker (Shakeel). They negotiated reasonable charges for quality services and mutual trust was built between the service-providers and owners. These links also save time when services are required urgently. Collective action has increased the owners' bargaining ability, self-confidence and the motivation of the group, while collective savings made the services affordable. The group's efforts have been essential to ensure affordable, timely and quality services to the horses and donkeys in their village, not only to improve welfare in the short term but also to assure sustained improvement in the longer term.

Deciding where to work and who to work with: targeting the neediest working animals

However motivated you are, it may not be practical for you or your organization to work with all animal-owning communities in your area. Villages may be geographically scattered in rural areas with few animals in each one, or the population of working animals in peri-urban or urban areas might be too large to cover all at once. Very few organizations have endless funds available, and it is important to target the animals in most need first.

Gathering information about your project area

This information can be collected by direct observation and by talking with animal owners and other community members. It can also be found in district and local government census records, animal health records, reports, surveys, stories, journals, maps and any other useful sources at national and local level.

Useful information to help you to decide which animals are at high risk for poor welfare may include:

About the area

- Number of districts, blocks or woredas in your project area
- Number and name of towns and villages in each block where working animals are kept
- Distance between these places and your project office
- Number of men, women and children in the villages
- Infrastructure and facilities available at different levels (district, block, village)
- Development organizations, private and government agencies already working in the area which may be interested in your work

About the animals

- Number and type of working animals in the area and how they are distributed (dense or scattered)

- Places where working animals congregate for work or to be traded, such as markets, brick kilns and factories

- Types of work carried out by animals in the area

- Their owners' livelihoods, income sources, economic and social status

- Opportunities and facilities available to improve animal welfare

During this process it may be possible to find out about the more common or severe welfare problems already known to be facing working animals, their nature and some of the possible causes or risks underlying them.

It always helps to prepare a simple format for these surveys and observations in advance, so that information can be recorded easily in the field.

Targeting the neediest animals

To discover where working animals are in greatest need, we recommend that you carry out a targeting exercise with a variety of stakeholders in the region or district. This exercise is based on the information you have gathered as described above, along with the inputs and experience of as many other people as possible, such as animal owners, users, carers and service providers.

Process box 2 below shows the steps for carrying out a targeting exercise. Case Study C illustrates a real example in more detail.

Some animals may do risky, difficult or particularly strenuous work but stay healthy and happy, because they are fit for the job and well managed by their owners. The targeting exercise is a good way for you to prioritise which groups of animals and owners to work with first. If you find out later that some of your early assumptions about animal welfare are not well-founded, you can adjust your priorities and plans in consultation with the animal-owning community.

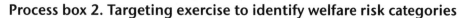

Process box 2. Targeting exercise to identify welfare risk categories

The first part of this exercise is designed to identify all the different groups of working animals in your region or district, initially based on the types of work they do. Brainstorm with as many stakeholders as possible, to identify discrete groups of working animals in the area.

From the information you have collected about the area and the animals, and any other sources available (such as other projects or publications), decide together on the criteria you will use in order to identify groups of animals at risk of the poorest welfare. Some examples might include:

- size and type of load carried on pack or pulled on cart;
- number of hours the animals work;
- distance travelled each day;
- external environment such as climate or terrain;
- owners' awareness of animal welfare;
- type of livelihood system in which the animal is used;
- economic pressures for using the animal;
- type of equipment used;
- any animal-based welfare information which you already have, such as the amount of lameness, wounds, disease or fearfulness among the animals;
- seasonal variations.

On a large piece of chart paper, draw a long horizontal line and mark one end 'High welfare risk' and the other 'Low welfare risk'. Show the groups of working animals along the scale, by agreeing on the relative risk of each group of animals having welfare problems, compared to the others. Groups of animals doing the same broad type of work, such as carrying packs, may have different levels of risk to welfare. For example, pack donkeys used in a small farming system, carrying vegetables or grain to the market once a week, may be at lower risk for welfare problems than pack donkeys carrying stones from the quarry to building sites on a daily basis. You may choose to place the farm donkeys nearer to the 'Low welfare risk' end of the scale and the stone-carrying donkeys nearer to the 'High welfare risk' end. Continue until all groups of animals are placed on the scale.

List the reasons for your decisions on the same piece of chart paper or record them separately, for future reference.

Divide your scale into three parts: high risk groups of animals, medium risk and low risk.

Case study C. Defining the neediest working animals for a pilot project in Ethiopia

Source: Lisa van Dijk and Brooke Ethiopia, annual planning workshop, July 2008, Addis Ababa, Ethiopia

Ethiopia has over 5 million working animals and most are donkeys used in rural transport and agriculture. A pilot project was initiated by the Brooke in 2008, covering several zones in the Southern National Nationalities and People's Region. Soon after the project teams started to gather information about working animals they faced a dilemma: the number of animals was just too big to reach effectively. For example, in Hadiya zone alone there were over 90,000 working donkeys.

In a planning workshop, the teams focused resources and effort on working with the animals which needed most help, and discussed how these animals would be identified. The workshop included representatives of the animal owners in each woreda. One group of participants started to define the animals working in Shashego woreda and their risks for having poor welfare. First they listed in detail the different groups of animals: pack horses and mules carrying grain to the market, gharry (carriage) horses transporting people, cart donkeys carrying stones from the quarry, cart donkeys carrying water to sell, and female pack donkeys carrying cattle fodder for the homestead. The first list was specific to Shashego and a second list was made for Lemmo.

Then the group asked themselves what factors they could use to decide which animals needed their help the most. They came up with the following list:

- work type, such as transport of goods or people by cart, pack animal, riding animal;
- workload, defined by the group as distance + weight + type of load + duration;
- working environment, including quality of roads and whether the work involved steep paths;
- whether the animal was used by the owner (and family) or hired out;
- whether the animal was working in an urban, peri-urban or rural area;
- owners' source of income: whether working animals provide the primary source of income or whether their owners have other income sources, such as farming;
- number of animals living or working in specific areas;
- number of animals per household;
- effect of the season on the animal and its work;
- number of people depending on the animal, in other words the family size;
- state of the equipment used, such as the gharry (carriage) and saddle;
- the results of a working equine welfare assessment previously carried out in some parts of the woreda.

They wrote each animal group on a card and drew a line on the floor to represent the risk of poor welfare, from low risk on the left to high risk on the right. A discussion followed on where each card fitted the scale, based on the factors above (see figure below). Decisions were based on the personal experience of group members combined with evidence from more formal assessments of animal welfare in the woreda. Participants sorted the animals into three main categories of need: high welfare risk, medium risk and low risk. In Lemmo woreda the animals categorised at high risk were cart donkeys and mules, the horses and mules at the grain market, and cart donkeys carrying stones and water. Animals at medium risk were homestead cart donkeys and garbage-collecting donkeys. The low risk animals were breeding and riding horses, the horse ambulances, and homestead pack donkeys.

This exercise provided Brooke Ethiopia with a basis for agreement on which groups of animals to target with their welfare improvement projects and how this could be done.

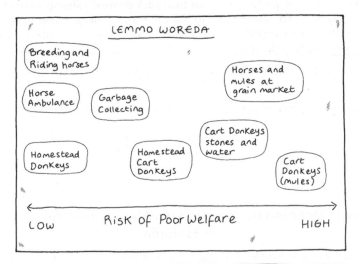

Figure 3.4 Identification of animal groups at risk of poor welfare in Lemmo and Shasego woredas

Deciding how to work: the intervention approach

Now you have identified all the groups of working animals in your project area and divided them into three categories: those at high risk for welfare problems, those at medium risk and those at low risk. The high risk category is likely to contain the neediest animals – the ones that your project may wish to focus on as a priority. This may result in working with fewer animals than if you used the same resources to tackle less severe welfare issues, so the priorities and strategic direction of your organization should be clear when choosing to target the working animals in greatest need.

Using these risk categories as a guide, we have identified three levels of intensity of engagement with animal-owning communities in order to improve the welfare of their working animals. The intensity of involvement with a community depends on the level of risk for poor welfare and the livelihood vulnerability of the animal owners.

The intensive approach

The intensive approach is the most effective way to make major improvements in animal welfare. It is suited to situations where the risks to welfare are high, combined with a high animal density and high vulnerability of the socio-economic or livelihood situation of their owners. One example is the owners and animals working in construction industries, such as brick kilns in India and Egypt. The intensive approach requires you, as the community facilitator, to meet frequently and directly with the community, facilitating them through a process of group formation and collective action to improve welfare. With your support, members of the group work together to:

- improve their own animal husbandry and management and their understanding of animal welfare

- improve the quality, accessibility and availability of existing service providers in their area

The core of this approach is to build the capacity of the animal-owning community to act as a collective for sustainable improvement in the welfare of their working animals. It requires time, effort and commitment from you and the community, although it can show excellent results in return. The steps in this process are described in detail in Chapter 4.

The extensive approach

The extensive approach is used where there are limits or constraints on the ability of you or your organization to work intensively with communities. These may be shortage of time and resources, or difficulty in accessing animals and their owners. This approach involves influencing the target group indirectly, by incorporating animal welfare improvement messages into the work of existing organizations in the area, such as women's groups, religious groups, unions or schools. It also uses mass media, like radio, posters or billboards, to convey animal welfare messages. The extensive approach may be used in areas with high or low animal density. However, welfare messaging is less effective than group formation in changing people's behaviour towards their animals, so it is most suitable for situations where the risks to animal welfare and the livelihood vulnerability of owners are lower. We find that these are usually populations of animals used for less intensive farming or domestic tasks and often located in rural areas. Strategies for the extensive approach are described in more detail in Chapter 5.

The semi-intensive approach

In reality, you will often find that you are unable to work intensively with all of the high-risk animals at the same time, either because your organization does not have the capacity or the animal owners are too scattered to initiate group formation for collective action. The semi-intensive approach is an intermediate approach. It is based on making maximum use of the intensive work that you are doing, by extending some of its effect to animals at high risk for poor welfare living in villages near to your intensive engagement groups. In this approach you meet communities directly as often as you can, but visits will be less frequent than to your intensively engaged villages. The principle of the semi-intensive approach is to create opportunities for cross-learning between animal owners in intensive and semi-intensive groups, such as village-to-village visits and competitions. In addition, semi-intensive communities are *linked with service providers* with whom you are already working in the nearby intensive communities.

Notes on the intervention approach

Intensity of engagement is <u>not</u> determined geographically. In one geographical area you can work with different groups of working animals and owners using different intensities of engagement. The approaches can complement each other and are not mutually exclusive. The decision on whether and how to work with a group of animal owners should be based on your best judgement and the strategic direction and capacity of your organization.

These approaches are based on our experience across several countries and in many environments and livelihoods contexts. However, there will always be exceptions. For example, a change in local government policy or its implementation following a mass media campaign may have a significant impact on the welfare of animals in high risk groups. In this case, an extensive, indirect intervention has led to improvement in the welfare of high risk animals.

Despite exceptions, the general rule is that if you want to make the maximum welfare improvement to the animals in most need (or at highest risk of poor welfare), you should try to work intensively with their owners as described above and in Chapter 4.

CHAPTER 4
Facilitation for collective action

What you will find in this chapter

This chapter contains the process for facilitating collective action by animal-owning communities in order to improve the welfare of their working animals. The six phases of this participatory process are summarised, along with their sub-steps. This is followed by a detailed description of each step, its purpose and the recommended process to follow. The new or adapted participatory rural appraisal (PRA) tools which may be useful at each step are signposted. A full explanation of these tools can be found in the Participatory Action Tools for Animal Welfare Toolkit (Part III).

It is not necessary to use all of the suggested PRA tools for each step. Using too many tools or exercises at the beginning may create confusion and bad feeling, which results in gradual loss of interest, low attendance and low participation in meetings. Choose your tools carefully. Remember that your goal as a facilitator is not the completion of particular tools in a particular way, or in a specified order. Your experience of community mobilisation for change will help you to decide which tool to use with the community to stimulate discussion and analysis, depending upon the specific purpose, to create a climate for collective action. If you have little or no previous experience, you should read the introductory materials listed in the references and further reading at the back of the manual, attend a training course and work with a more experienced community facilitator before starting the facilitation process yourself.

The chapter is organized as a practical manual. If you are a less experienced facilitator you can work through the steps in sequence with the community. An experienced facilitator may prefer to adapt the process to his or her own experience and incorporate specific PRA tools for animal welfare change in a more flexible way.

An outline of the whole process of collective action for welfare improvement is shown in the figure below, with each phase and its detailed steps described in the rest of this chapter. The case study on page 67 gives you an overview of the whole process carried out with a real community group.

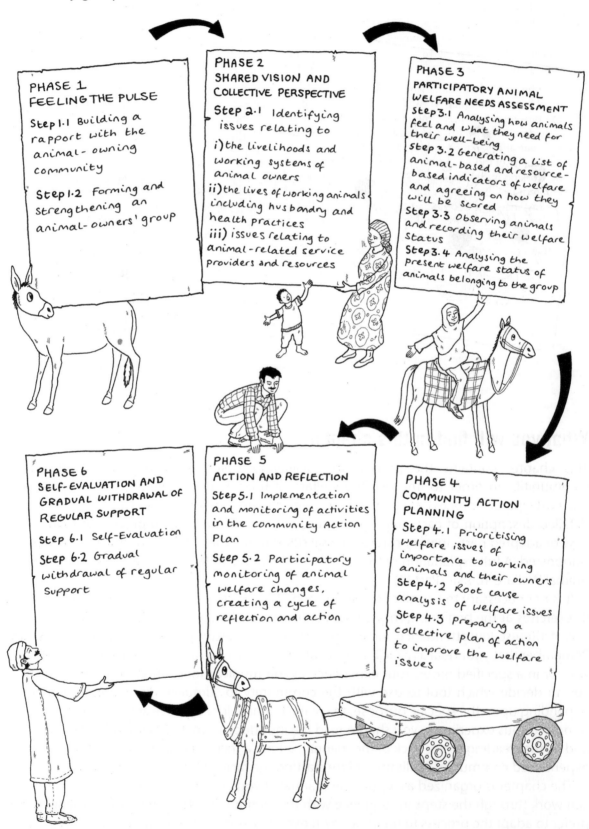

PHASE 1
FEELING THE PULSE

Step 1.1 Building a rapport with the animal-owning community

Step 1.2 Forming and strengthening an animal-owners' group

PHASE 2
SHARED VISION AND COLLECTIVE PERSPECTIVE

Step 2.1 Identifying issues relating to

i) the livelihoods and working systems of animal owners
ii) the lives of working animals including husbandry and health practices
iii) issues relating to animal-related service providers and resources

PHASE 3
PARTICIPATORY ANIMAL WELFARE NEEDS ASSESSMENT

Step 3.1 Analysing how animals feel and what they need for their well-being
Step 3.2 Generating a list of animal-based and resource-based indicators of welfare and agreeing on how they will be scored
Step 3.3 Observing animals and recording their welfare status
Step 3.4 Analysing the present welfare status of animals belonging to the group

PHASE 6
SELF-EVALUATION AND GRADUAL WITHDRAWAL OF REGULAR SUPPORT

Step 6.1 Self-Evaluation
Step 6.2 Gradual withdrawal of regular support

PHASE 5
ACTION AND REFLECTION

Step 5.1 Implementation and monitoring of activities in the Community Action Plan

Step 5.2 Participatory monitoring of animal welfare changes, creating a cycle of reflection and action

PHASE 4
COMMUNITY ACTION PLANNING

Step 4.1 Prioritising welfare issues of importance to working animals and their owners
Step 4.2 Root cause analysis of welfare issues
Step 4.3 Preparing a collective plan of action to improve the welfare issues

Table 4.1 Overview of the process of facilitation for collective action to improve the welfare of working animals

Case study D. Collective action for improvement of working animal welfare

Source: Anup Singh and Sishupal Singh, Brooke India, Baghpat, Uttar Pradesh, October 2009

In June 2007, Brooke India's Baghpat district equine welfare unit started to work in the village of Galehta in a remote area of Baghpat district, which has about fourteen working horses and six mules owned by a community of potters. The Brooke team's field facilitator, Sishupal Singh, visited the village several times to meet animal owners, the village head and other representatives of the community. A general meeting was held on the 18th of June with all the animal owners, and during this meeting everyone made a commitment to meet regularly every month to analyse their own situation and that of their animals.

First the owners conducted a Mapping exercise (Toolkit T1) where they made a map of their community, identifying the animals in each household and the important community assets in the village, including the tyre repairing shop, medical store, school, post office and the locations of animal-related service-providers. The group used Seasonal analysis (T6) to explore the availability of their labour throughout the year and variations in workload, flow of income, occurrence of common animal diseases and the availability of fodder and feed. November to June was the busiest period, when they earned extra money by working in the brick kilns. Sishupal then introduced the group to a Venn diagram (T3) to analyse the different facilities used in their daily lives, such as the hospital, market, animal feed seller and farrier, and the distance that people had to travel to these, how often they were used and their importance to the villagers. An in-depth analysis of animal diseases was carried out using Matrix ranking (T9) including diseases affecting horses and mules and exploring how their owners get access to services such as veterinary and other animal health providers.

After this period the group had a clear common understanding of the issues which directly or indirectly affected their animals' welfare and their own livelihoods. Some actions were agreed. For example, one day during a group meeting in the first three months of their work together, an owner asked about a disease called *'hiran bayal'* which he had seen in a horse from his relative's village. Sishupal asked about symptoms of the disease, which was identified by the group as tetanus. The group discussed possible implications of this disease if it happened in their own village and used the Animal welfare cost-benefit analysis tool (T15) to talk about how they could prevent their animals from getting tetanus. Sishupal recognized that this was a good opportunity to initiate an entry point activity through collective action to vaccinate all the horses and mules in Galehta against tetanus. Based on their animal welfare cost-benefit analysis the owners took action together. They all contributed money, bought the vaccines from a medical shop and called the local veterinary service provider to vaccinate all of their animals. Everyone realised their collective strength and they were enthusiastic to go ahead with other kinds of collective action for their animals' benefit.

The discussions on the collective contribution of money led to an analysis of the saving and credit opportunities available to group members. During a meeting in January 2008 Sishupal introduced a Credit analysis exercise (T13), through which the owners gained a greater understanding of the losses and benefits they experienced through their existing credit system and the impact that these had on their social status and their animals. Many of them had taken loans from local money lenders, at interest rates of about 5% per month, or 60% per annum. One horse owner, Ilam Chandra, suggested that they form their own self-help group to meet their needs for credit to provide for animals and their own families. Seventeen owners formed a self-help group and each member contributed 100 Indian rupees (Rs.) per month.

The women, who always participated in the village meetings, became interested in forming a women's self-help group. After a few meetings with the men's group members, the women formed their own group with eleven members and started collective savings of Rs. 50 each per month.

As part of their ongoing situational analysis, the men's group went on to look at how they accessed the services of farriers, hair-clippers and cart-repairers. They used the Venn diagram which they had made in a previous meeting, along with Mobility mapping (T2) and Dependency analysis (T12), to analyse all the benefits and challenges of using these service providers. Dependency analysis helped the group to understand the extent of their dependency on each service. This led the group to invite the preferred service providers to their meetings, in order discuss the problems they faced and negotiate better services. Sishupal invited the veterinary doctor to join the group during one of the meetings and help them to discuss diseases such as colic (abdominal pain) and surra (*trypanosomiasis*) in more detail. They used Three pile sorting (T23) to look at the symptoms of each disease, the cause of disease and any preventive measures which could be taken against each one.

By March, the group had experience of successful collective action and though their use of animal-adapted PRA tools the members had a better understanding of their own situation. Sishupal initiated more in-depth analysis of the present animal welfare situation in Galehta, using If I were a horse (T17).

He asked the participants 'If you were a horse, what would you expect from your owner?' The group listed a range of 14 items. Sishupal then asked 'So how far are your horses' expectations fulfilled?' based on their present management practices and according to how much they could afford or make available for their animal. Everyone then discussed reasons for the areas where scores were low.

In their following meeting, in April, Sishupal returned to the If I were a horse exercise, asking 'If the expectations of your horses were not fulfilled, what would be the effect on them?' The group came up with effects such as dehydration, weakness, insects on the skin, a dirty body and no power for work. Then Sishupal asked 'How would you see these on the animal?' Participants listed sixteen different physical and behavioural signs on the animal, including wounds on the back, withers and belly, hoof cracks, swelling of joints, visibility of bones and ribs and degree of alertness.

The following month Sishupal asked the group how they would score or rate their animals based on the physical and behavioural signs or parameters that they had developed in the previous meeting. A big discussion followed, in which each owner felt that his own animal was much better than his neighbours'. Rajpal became so annoyed that he suggested they go and have a look at all the horses and mules in the village. The group decided to visit each animal and score them according to the parameters they had developed. They added five more, relating to the use of service-providers and resources identified during the situational analysis. Together they agreed on a traffic-light scoring system for each of the 21 welfare indicators, using red for poor, orange for medium and green for good.

The group did their first Animal welfare transect walk (T22) over a period of three days. A lot of welfare problems were detected and recorded during this time: dirty eyes, wounds on various parts of the animals' bodies, swelling of knee joints, fearful animals, dirty skin, dirty and foul-smelling hooves, poor quality saddles, signs of firing (branding with hot irons) and badly shaped horseshoes. Taking part in this exercise led some of the owners to realize that their animals were in poor condition and they took immediate action to improve this.

In the meeting that followed, the traffic light chart was analysed and Sishupal initiated an Animal welfare cause and effect diagram (T26) to identify the root causes of the welfare issues, one after another. This exercise helped all of the group members to recognize the seriousness and the cause of each issue. For example, they realized that cart balance and overloading were contributing to wounds on different parts of the body. They decided to develop an action plan based on the situational analysis and root cause analysis for their priority animal welfare problems. For each welfare issue affecting their horses and mules, owners identified the activities that would be carried out, who would be responsible for doing them and who would monitor them, as well as a time-line for doing this. They also agreed to record everything they did properly on chart paper. After three months of implementing their action plan, the group repeated the Animal welfare transect walk with Sishupal and recorded their findings on the traffic light chart. They analysed the chart and adjusted their action plan according to the findings.

In August 2009, one year after forming their animal welfare action plan, another meeting was organized by the group members to review the past year. They used the Before-and-after analysis exercise (T11) to understand the changes that they had made, and identify areas for further improvement in the village action plan. They also compared the Animal welfare transect walk traffic light charts which they had filled out on four occasions. Looking at the changes over time showed that they had brought about a significant improvement in animal welfare. Overloading was fully controlled, there were fewer wounds despite a hard season of work and the occurrence of disease was reduced to a great extent. They also identified that the welfare of their animals had not been stable: over the year it was influenced by the season because specific causal factors occurred at particular times of year. This information enabled them to plan ahead for the future. As well as positive changes in the lives of their animals, the group recognized that animal welfare improvement had had a beneficial effect on the lives of all the animal owners and their families in Galehta.

In their October meeting the group of owners decided to hold a regular competition to reward the owner whose animal's welfare status had improved the most. They discussed this with Sishupal and refined the monitoring system to use numerical scores for each welfare issue, instead of the traffic light signals. This enabled them to measure smaller changes in their animals' physical and behavioural signs or indicators, and to agree at each meeting which member had made the most improvement. With a regular chance to increase their own status in the village by winning the competition, animal owners worked even harder to implement their action plan and reduce any signs of poor welfare on their horses and mules.

By this stage the Galehta self-help group was making its own decisions and seeing positive changes in animal welfare at every meeting. Together the participants identified what would need to be in place for them to be able to continue without Sishupal's help, and the indicators which would tell them when they were ready. Using these indicators they have prepared a plan of action to enable Sishupal to withdraw his regular support. Now, by mutual agreement, the Baghpat district equine welfare team's visits will be gradually reduced, until Sishupal only goes to Galehta when he is really needed and just occasionally, by invitation of the group, to join a monthly meeting and celebrate success.

How long does the process take and when will you start to see improvement in the welfare of the animals? We have seen small positive changes in welfare right from the start of Phase 1 'Feeling the pulse' and Phase 2 'Shared vision and collective perspective'. More substantial improvements can be seen from Phase 3 'Participatory animal welfare needs assessment' onwards. In this phase the group starts to assess the welfare of their own working animals and develops concrete action plans for change. This is a direct result of your capacity-building of the group the first two Phases, so it is essential to stick with the process in the early stages – even if changes appear to be small – in order to reap much larger benefits for the animals later on.

Having facilitated this process with many different animal-owning groups and in different contexts, we estimate that it will take about 36 months to go through the whole process with a particular community group (although of course you may be working with several groups simultaneously during this time). Phase 1 'Feeling the pulse', and Phase 2 'Shared vision and collective perspective' will take approximately six months. The group will have developed their community action plan as part of Phase 3 by the end of the first year. In the following 18 months (up to abut 30 months of intervention) the group will build its capacity through the implementation of the action plan and through the 'Action and reflection' cycle (Phase 4). A progress review can take place midway, after 18 months of intervention. The group should be ready for you to withdraw your intensive support after about 36 months, leaving the community able to generate and sustain improvements in animal welfare by themselves.

Phase 1. Feeling the pulse

We have called the first phase of facilitation for collective action 'Feeling the pulse'. The purpose of this phase is to understand the community better, gain trust in each other and create an atmosphere ready for change. This phase brings the community together around a common goal and builds their confidence in their own ability to bring about positive change in their animals' lives by working together as a group. The steps which make up Phase 1 are explained in detail below.

Step 1.1: Building a rapport with the animal-owning community

The first part of feeling the pulse is getting to know the animal-owning community.

- Introduce yourself as a field worker from an organization that is interested in supporting and organizing community-based groups to work towards sustainable improvement in animal welfare.

- Ask people about their lives, their problems, local culture and habits.

- Visit village shops and meeting places for informal discussions.

- Hold meetings with village leaders and talk with all interested individuals, including school teachers, religious leaders and anyone else who can support you from the beginning to organize the community.

- Strengthen contact with the families who keep animals, including women and children who may be responsible for managing animals at home.

- Identify and talk with local veterinary and animal health service providers, medicine shop keepers (agrovets), farriers and anyone else who works with animals, directly or indirectly.

Phase 1 Feeling the pulse

Step 1.1 Building a rapport with the animal-owning community

Purpose
- To build confidence and mutual trust
- To initiate the process of group formation

Process
- Get to know the community
- Carry out entry point activities to initiate group formation

Tools (see toolkit)
- Mapping (T1)
- Daily activity schedule (T4)
- Gender activity analysis (T5)
- Historical timeline (T7)
- Animal welfare snakes and ladders game (T16)

Step 1.2 Forming and strengthening an animal owners' group

Purpose
- To organize owners and users into groups
- To strengthen the groups and local institutions

Process
- Identify and form a local community group
- Stabilize and strengthen the group

Tools (see toolkit)
- Seasonal analysis (T6)
- Dependency analysis (T12)
- Credit analysis (T13)

Table 4.2 Process overview Phase 1: Feeling the pulse

Taking part in daily or regular activities with people, or taking part in important events at the invitation of the community, such as ceremonies, funerals or celebrations, can help you to understand the community and bring you closer. Before beginning to guide animal owners or users towards any specific welfare intervention, you should try to get a feel for the important issues which might have far-reaching effects on the welfare of their animals. Collective action does not happen spontaneously, it is triggered by a pressing need to act in the face of problems or crises.

Your ability to understand the needs and hopes of community members can help you to play the role of a catalyst for change. Some of the pressing issues in the lives of animal owners may not have a direct effect on the welfare of their animals, but understanding these can enhance your acceptance as a supporter and well-wisher for the community.

WARNING BOX

During this period, which may take one to three months, it is important that no education or intervention programmes are conducted.

Why?

- An intervention started without knowledge of local problems and context is likely to be the wrong one for the community

- An intervention started without a genuine rapport with and understanding of the community is likely to be viewed with suspicion

- An intervention started without a plan for sustainability is likely to create dependency on you or your organization, which will be very hard to undo later.

Your initial interactions are your opportunity to:

Understand the lives of animal owners, users and carers

Which elements of the daily lives of animals or their owners, if changed, could improve their quality of life substantially? For example, do people and their animals have to walk a long distance to fetch water for the household? Are they obliged to work for an excessively long time to repay a loan?

Find out how much interest and motivation is present in the community to improve their animals' welfare

- How keen are the people to seek a change?

- Does the community have a critical mass of such people?

- Are the interested people influential enough to bring other animal owners along with them to seek the same change?

- Are people willing to start acting of their own accord and relying on themselves to achieve the changes they would like to see?

- Do they understand the costs and demands of taking action?

Understand the role of working animals

This period is your opportunity to understand the role played by working animals in the lives of their owners and families.

Identify social groups which may be interested in animal welfare

Your initial interactions are also your opportunity to identify social groups and interest groups within the community (see Figure 4.1). If there are existing groups, organizations or unions, it may be possible for them to make a collective decision to include animal welfare in their activities within their regular meetings.

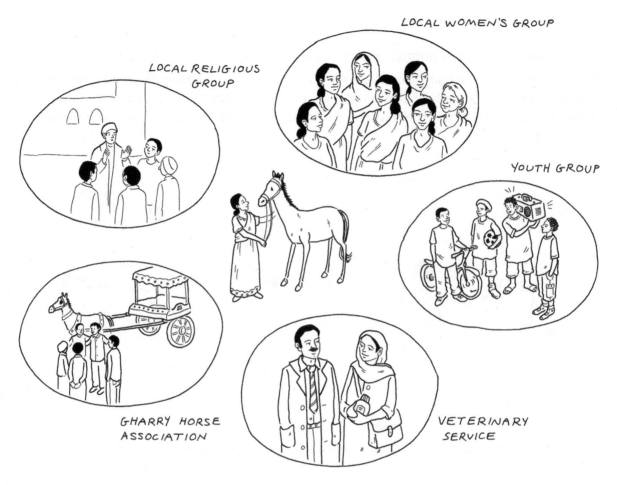

Figure 4.1 Different social groups and interest groups within the community

Entry point activities to initiate group formation

Next, discuss and identify activities which specifically involve the people who own or are interested in animals. By sorting out one or more problems through collective action, these 'entry point activities' help the community to realize their collective strength. They bring mutual trust and confidence in each other and can start to raise awareness about animal welfare. It is more important to ensure participation and decision-making by the group than to select a welfare-related activity, if this is not people's primary interest at this early stage.

Some entry point activities which we have used successfully:

- Repairing the road into the village, so that working animals can pull their carts more easily.

- Replacing sharp animal-tying stakes with soft rubber tyres to prevent injuries (see Case study E on page 76).

- Community savings groups that benefit working animals – we find this to be one of the most effective ways to bind people together in a group working for a common purpose, which lasts without the continuing support of an external organization (see Case study F on page 78).

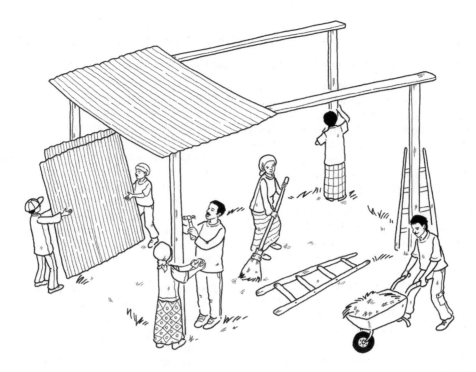

- Exposure visits, to see how successful community groups organize themselves and tackle a problem or transform a situation, can be very valuable at this stage of group formation. It is important that you facilitate as much learning as possible from an exposure visit, but do not direct people to find similar solutions for their own location. Ask your group to consider not only what is done by the other community, but who does it and how it is done. For example:

  o What activities do the existing group carry out collectively? How do they do them?

  o If the visit is to a group with its own savings fund, how was this decided upon and how are contributions collected?

  o Who takes responsibility for negotiating with outside agencies?

  o Who maintains the records for the group?

Case study E. Entry-point activity with an animal-owning community

Source: Ramesh Ranjan, Brooke India, Basantpur Sainthli village, Ghaziabad, India, October 2007

The Brooke India Ghaziabad district unit equine welfare team works with groups of animal owners throughout Ghaziabad district in Uttar Pradesh. Every year the team expands its reach to new villages. To build rapport with the animal owners in a new village, they start with an entry point activity based on "appreciative planning and action" (APA) methodology (see box).

In October 2007 a community facilitator visited Basantpur Sainthli village, which has 24 working animals, mainly horses used in brick factories (kilns) and for transport of goods and people. After initial orientation, he organized a meeting with animal owners in the village and started discussion by asking about any positive actions which had been taken collectively by the group. They explained proudly that they had lobbied district administrative officers to stop illegal motor traffic on local roads. It had been a very successful joint effort. This was the Discovery step of APA.

The facilitator then asked what challenges faced the owners when taking care of their animals. Various issues were brought up, one of which was the stakes used to tie up animals. Horses were seriously injured by lying on the stakes and could not be worked for several days. It cost a lot of money to get treatment from a nearby village.

The group was encouraged to prioritise the issues identified and to select one which could be sorted out collectively without the help of outsiders. The stake issue was agreed as the one that was causing many problems and could be easily solved. The facilitator asked the group to think of all the possible benefits of solving this issue. The group came up with many positive benefits for their animals, such as prevention of injury, and also for themselves, such as reduction of cost of treatment. This was the Dream step.

By now the group's enthusiasm was high and they agreed to sort out the issue immediately, replacing the wooden stakes with half-buried bicycle tyres, because several owners had seen this in another village and knew that tyres would not harm the animals. This was an easy solution, because bicycle tyres were easily available in their own village (Design step). Together they went from one house to another, helping each other to remove the stakes and replace them with bicycle tyres. Within two hours the group had solved one of their concerns (Delivery step).

This collective action built a positive feeling within the group and the owners decided to come together more often to take action on issues affecting their working animals.

Appreciative Planning and Action (APA) is a framework which helps to empower groups and communities to take positive action for their own development. It is built on the principles of searching for positive events, for successes, for what works, and for what gives energy to individuals and groups. It then seeks to empower local communities to take action by creating a vision of an even better future, making commitments, and then taking the first step straight away. There are four steps involved in this process:

Discovery: ask positive questions, seeking what works and what empowers.
Dream: envision what could be and where we want to go.
Design: make an action plan based on what we can do for ourselves, making personal commitments.
Delivery: start taking action now.

Source: Bhatia, A., Sen, C.K., Pandey, G. and Amtzis, J. (Eds). *(1998) Appreciative Action and Planning, In: Capacity-Building In Participatory Upland Watershed Planning, Monitoring And Evaluation – a resource kit.* ICIMOD, Kathmandu, Nepal, pp. 127–133.

Step 1.2: Forming and strengthening an animal-owners' group

At this point you have found a number of people who are interested in the same thing (working animals) and they have taken part in an activity together. Developing these people into a functioning animal owners' group is a crucial step which needs your support and experience. The group can be composed of men, women and children, or separate groups may be formed as appropriate. We find that where it is difficult to initiate collective action by men, organizing women's groups is easier and the men come together after seeing the success of the women. In some cases it may be necessary to meet with the men before forming women's groups, in order to explain what is going on and the benefits that a women's group will bring to the family.

It takes several meetings over a period of two or three months for the group to have a definite membership and up to a year for a strong and stable group to be established. In the meetings held during this period the members will also raise issues concerning the family and the community. If the group involves common contributions, savings and lending, these are debated thoroughly (see Case study F on page 78). The willingness to abide by group decisions without breaking agreements and confidence in the group indicates the degree of trust that members have in one another.

Encourage the group to frame and review their own set of norms, rules or regulations. We find that these usually include:

- **Membership**: who may join the group, what the group size should be and what happens when a member leaves

- **Meetings**: what is a quorum for the group, what happens if people are absent from meetings or turn up late to meetings

- **Representatives**: who represents the group, how representatives are chosen, whether they are rotated periodically and if so, how often this occurs

- **Sanctions**: what sanctions are needed for violation of the rules, and when exceptions may be made

- **Common contribution**: if membership of the group involves a common contribution or savings fund, what is the minimum amount to be contributed, whether withdrawal of savings is permitted, whether and how interest is paid on savings

- **Loans:** if loans can be taken from the savings fund, how these are prioritised, what interest rate is charged, how the use of loans is monitored, and how defaulters are penalised for overdue loans

Group stabilisation and strengthening is an ongoing process. By sticking to the agreed rules or norms, making collective decisions on a common action plan and carrying out these actions either individually or together, the animal owners' group becomes increasingly strong and effective.

Case study F. Saharanpur credit and savings groups – empowering the rural poor to ensure the welfare of their working animals

Source: Kamalesh Guha and Dev Kandpal, Brooke India, Saharanpur district, Uttar Pradesh (2008)

In Saharanpur district of Uttar Pradesh, donkey-owning communities suffer from cycles of seasonal employment and very low wages. While interacting with workers in the villages, it became clear that lack of unity among them was a major hindrance to solving their common problems. The major problems experienced were exploitative money-lenders, competition between mechanised and traditional forms of transport, poor road conditions, marginalisation of donkey owners and low self esteem. Animal owners were encouraged to form Self-Help Groups (SHGs): member-managed collectives of typically 10 to 20 men or women who create a common fund through regular small savings and provide financial services to their members. Convinced by the concept of establishing SHGs, donkey owners from 16 villages came forward. Gradually, with the assistance of the District Co-operative Bank, they opened bank accounts, laid down bye-laws defining the role and rights of each member of the group and decided the action to be taken against any member who violated these agreed terms and conditions.

These groups meet every month to discuss their common issues and donkey welfare is high on their agenda. Group members have started to take loans from their common funds mainly on the basis of mutual trust, with minimal documentation and without any tangible security. Usually small amounts are taken for short durations and used mainly to help their animals (see box). Earlier these owners had to pay an exorbitant rate of interest to local money-lenders, as high as 60–120% annually. Now SHGs play an important role in maintaining funds for veterinary treatment and for taking preventative action, such as vaccination against important diseases. Members of existing SHGs are assisting new groups to form in the same and neighbouring villages. The SHGs have empowered women significantly, as several of these groups are managed by women who play a major role in providing care to working animals. The ability of these groups to continue to function and grow without any external help is a hope for long term sustainability of animal welfare as well as the well-being of donkey owning families.

> **Details of group savings and use of loans for 16 Self-help Groups from January to June 2008**
>
> Total number of donkeys benefited: 141
> Total number of families benefited: 199*
> Total savings of 16 SHGs: 183,500 Indian rupees (Rs.)
>
> *Total amount of loans taken by group members:*
> For purchase of animal feed: Rs. 47,700
> For purchase of animals: Rs. 43,500
> For repairing carts and other maintenance: Rs. 39,000
> For veterinary treatment: Rs. 6,000
> For repaying earlier loans to buy animals: Rs. 24,000
> For other domestic needs: Rs. 42,000
>
> ** Note: this includes some families who do not currently possess an animal but belong to same community*

Process box 3. Critical features of a local community group – indicators after a year

An owners' group is not just any collection of animal owners, handlers and carers. It is those who come together with commitment to improving animal welfare, with a sense of direction and a plan for the future. The group must have the following characteristics:

The members decide on plans and take collective action to carry them out.

The group meets regularly at an agreed place and time, and its members participate actively.

Attendance Sheet				
Name	19/7	24/8	15/9	20/10
Zeny	✓	✗	✓	✓
Tesfaye	✓	✓	✓	✓
Beruk	✗	✓	✓	✗
Michael	✓	✓	✓	✗
Dibora	✓	✓	✓	✓
Simon	✓	✓	✓	✗
Tewodros	✓	✗	✓	✓
Yodit	✓	✓	✓	✓
Solomon	✓	✓	✓	✓
Senait	✓	✗	✓	✓
Dagim	✓	✓	✓	✓
Surafel	✗	✓	✓	✓
Kaleab	✓	✓	✓	✓
Emebet	✓	✓	✗	✓
Alula	✓	✓	✓	✓
Kidus	✓	✓	✓	✓

There is free and open communication and feedback among the members.

CRITICAL FEATURES OF A LOCAL COMMUNITY GROUP

The members know about membership criteria, rights and responsibilities. They have decided on the rules and procedures for meeting and working together. These rules may be formally written down, or they may simply be commonly understood.

The size is big enough to be effective, yet small enough to allow members to interact and participate.

They come together with a common interest to work towards a shared objective, goal or purpose. All members have the same understanding of their reasons for meeting.

There is an identified leadership. This may be an individual or more than one person, but the group recognizes the leadership, and the leaders lead actively.

Case study G. The Butajira Ghari Horse Owners' Association

Source: Gorfu Naty and Kibnesh Chala, Brooke Ethiopia, Addis Ababa, January 2010

The Butajira Ghari Horse Owners Association in the Southern Nations Nationalities and Peoples Region (SNNPR) of Ethiopia was established in February 2008. Its 670 members own over 1200 Ghari (taxi) horses, which provide transport services to the people of Butajira town and the nearby villages.

The Ghari owners were facing several problems with their animals, including lack of shelter, water and feed, health problems such as epizootic lymphangitis (EZL) and hoof problems caused by inappropriate shoeing. The side roads in Butajira were not maintained well, so a combination of big potholes and ill-fitting harnesses strained the horses and contributed to wounds. Ghari owners realized that even if they loaded up to seven people in a cart this did not guarantee more income, because passengers were not charged a fixed fare. In addition, Ghari drivers were often stopped by the traffic police, because the law prohibited horses from crossing the main asphalt road, even though there were no alternative roads to use.

According to Asres Berta, now the Chairman of the Butajira Ghari Horse Owners' Association, some individuals had contemplated forming an association to overcome these problems, but did not know how to start. Early in 2008, Brooke Ethiopia organized an animal welfare sensitisation workshop for Ghari horse owners, traffic police, and a representative from the agriculture office. This gave the owners insight and encouragement to form the Association. Initially many Ghari horse owners were reluctant to join, but after the benefits of the Association become more visible, more members joined in.

The main activities of the Association are currently:
- Setting control mechanisms to prevent Ghari horse owners from overloading their horses.
- Creating a savings and borrowing system within the Association.
- Addressing their common animal welfare problems collectively.
- Improving the feed supply for the horses.
- Improving the equipment they use, such as harness, horse shoes and cart tyres.

The Association has made significant progress in the last two years:
- They were able to implement a law in their Associations' constitution that limited Gharis to a maximum of three passengers (instead of up to seven). The Ghari fare was doubled from 50 cents to 1 Birr, so that owners would still make the income needed to look after their animals and families. Collectively they are trying to reduce overloading in town, by setting up an overloading control mechanism which fines offenders along the six Ghari transport routes.
- They have repaired the potholes in 17km of side roads which were making the trips difficult for the horses. Each association member dedicated four days to maintaining these roads.

- Their savings and borrowing scheme has removed the pressure on Ghari owners to work long hours in order to pay for the traditional weekly revolving fund. Members have also started to borrow money to buy additional horses so that their current horses can rest on alternate days.
- The Association opened a shop which now buys animal feed in bulk and sells it to members at cost, so they are protected from the price rises which previously reduced the amount of feed they could afford for their animals. The shop also makes water available for the animals while they are working.
- Members have linked up with local veterinary services. They have invited farriers to work in front of the Association's shop and negotiated better quality shoeing.
- The Association creates a forum for members to meet and discuss the welfare of their horses. They have begun to compete with one another in terms of how well their horses are looked after and give advice to members whose horses are in poor condition.

The Association has provided Ghari owners with a powerful platform to improve their horses' welfare, as well as making great contributions to the livelihoods of their owners.

Phase 2. Shared vision and collective perspective

The purpose of Phase 2 is to identify common animal welfare goals within the group which you have facilitated and strengthened in the first phase. You can do this through a series of steps which enable the group to analyse their own situation and that of their animals. Phase 2 identifies and analyses issues related to three areas as follows.

Phase 2 Shared vision and collective perspective

i) The livelihoods and working systems of animal owners
Purpose
- analyse group members' livelihoods and working systems
- identify issues which have a direct or indirect effect on animal welfare (such as livelihoods, income, debt, dependency and others)

ii) The lives of working animals
Purpose
- analyse the lives of working animals, their feed rest and work patterns, daily and seasonal variations
- identify issues related to animal diseases and their prevention (such as recognition of disease, seasonal disease patterns and methods of prevention)

iii) Animal-related service-providers and resources
Purpose
- analyse service-providers contributing to animal welfare and identify issues related to stakeholders' practice towards group members and their animals
- analyse resources needed by animals and to identify issues related to the availability and quality of resources

Process
- organize group meeting(s) to analyse issues relating to group members
- identify group members' issues which have a direct or indirect effect on animal welfare, through discussion, use of PRA tools and observation
- discuss and analyse these issues with different sub-groups, such as animal owners, handlers, carers, men, women and children
- present analysis to the larger group for wider agreement on issues identified; visit stakeholders and health service providers or visit sites and resources together with the group members; discuss welfare issues with identified stakeholders individually or by organizing a separate stakeholders' workshop; follow-up with a group meeting to analyse gaps in current service providers' practices and resources and to discuss options for improvement

Tools (i)
- Mapping (T1)
- Mobility map (T2)
- Venn diagram (T3)
- Daily activity schedule (T4)
- Gender activity analysis (T5)
- Gender access and control profile (T10)
- Seasonal analysis (T6)
- Changing trend analysis (T11)

Tools (ii)
- Animal welfare and disease mapping (T1)
- Animal disease venn diagram (T3)
- Daily activity schedule (T4)
- Dependency analysis (T12)
- Animal body mapping (T20)
- Animal welfare practice gap analysis (T21)

Tools (iii)
- Animal-related service and resource mapping (T1)
- Mobility mapping (T2)
- Pair-wise ranking (T8)
- Matrix scoring of animal-related service providers (T9)
- Cost-benefit analysis of animal-related service providers (T15)

Table 4.3 Process overview Phase 2: Shared vision and collective perspective

Livelihoods and working systems of animal owners

This includes understanding the composition of the community: who lives where, the number of animals in the household and the number of people dependent on an animal. It also looks at their daily activities, the difference in activities between family members, the family's main sources of income, their expenditure and credit requirements and any seasonal fluctuations in household income. There may be other relevant pieces of information to explore, for example people's dependency on other stakeholders to help care for their animal.

The lives of working animals

This refers to the daily activities of the animal, seasonal variations in feed and disease, the daily mobility (journeys) of the animal, including the load carried or pulled, distance covered and frequency of trips. Analysis of the group's current husbandry and health practices related to both prevention and treatment of their animals' welfare problems is also important.

Animal-related service-providers (local health providers, farriers, hair clippers, cart-makers, medical stores, feed sellers and others) *and resources* (feed, water, grazing, shoes, harness, medicines and others). This includes understanding the location of people who provide animal-related services, their distance from the community, their availability, affordability and quality and the preference or acceptability of each service provider to the community. Information about animal-related resources includes the cost and quality of different resources and their local and seasonal availability.

Situational analysis is carried out using the participatory methods and tools summarized in the table below.

It may take several visits and community group meetings to complete Phase 2. The number of meetings needed and the tools you will use to identify and analyse all the issues in this Phase are flexible, and will vary between groups. We find it best to facilitate these meetings through a schedule of regular visits until the analysis is completed.

All the exercises in this phase follow the same basic pattern:

- Organize a community meeting for analysis

- Using PRA tools and discussion, encourage the group to identify issues which have a direct or indirect effect on animal welfare.

- Analyse the issues with different sub-groups, such as animal owners, handlers and carers, or men, women and children.

- Present the sub-group analyses back to the wider group or community for further discussion and to plan a course of action.

You can also encourage the group to meet individual stakeholders and service providers at their places of work, or to hold stakeholder meetings or workshops.

Process box 4. Animal needs and human needs

People and working animals are dependent on each other to survive. They must 'Share the Load' in order to thrive. However, there are sometimes conflicts between the needs and feelings of animals and people: what is best for the family's income may not be best for the animal's welfare. This is normal when money and resources are limited and have to be shared out between meeting the needs of people and their working animals. There is great value to working with communities in order to find the balance for both animal and human welfare.

Especially when you start to work with a new group, the animal owners, users and carers will come up with needs and issues which do not seem to have a clear relationship with animal welfare, although they do affect the group members directly. Examples include low daily wages, high interest rates on loans, children's health and education, bad road conditions and limited living space. This is very normal and should not be discouraged. Your support agency may not have the scope to deal with all these issues directly; however some things can be done:

- The process of group formation and collective action for improvement of animal welfare also builds the self-reliance and capacity of the group to deal with other problems. The group becomes empowered to sort out their own problems collectively wherever possible, without waiting for external help. This might happen without any formal action plan.
- Initiating a credit and savings group not only benefits the animal (see Case study F on page 78), it also solves some of the issues which might not be directly related to animals, such as the high interest rate charged for loans from outside sources.
- You can also provide information or support community groups to contact other organizations or agencies in their locality which could help with issues such as health or education.

Figure 4.2 Balancing animal needs and human needs. (Adapted from *The Two Mules, a Fable for the Nations: Co-operation is better than conflict*, Society of Friends Peace Committee, Washington)

Phase 3. Participatory animal welfare needs assessment

The purpose of Phase 3 is to look at the present welfare status of working animals, by bringing the animal itself to the centre of the group's analysis. We call this 'Participatory welfare needs assessment' (PWNA). It is the most important phase of facilitating collective action to bring about real improvement in the welfare of animals and focuses on identifying the physical signs and behaviour related to physical and mental welfare. It takes four steps, using specific tools for each step, to reach the deeper levels of analysis which enable animal owners to see the world from their animals' point of view and understand their needs.

This is a very effective process for sensitizing owners to their animals' needs and feelings and the ways in which these are expressed through an animal's behaviour or body language. Being able to listen to the voice of the animal is crucial in creating motivation among owners, users and carers and enabling them to generate action plans and monitor animal welfare over time, so that improvements can be sustained in the long term after your inputs have reduced or stopped.

What you can do with participatory welfare needs assessment:

- Facilitate the group to identify and monitor the welfare status of any animal, by looking at the physical condition of its body and assessing how it feels through observation of its behaviour.

- Facilitate the group to identify the things which may affect the welfare of their animals. These include management practices and behaviour by animal owners, resources, stakeholders and services and the effects of the environment.

- Facilitate the group to assess the level or severity of various welfare problems and their contributing factors.

- Sensitise the group to the welfare needs of their animals, in other words the types of work, handling, management, environment and resources which will promote good welfare.

- Encourage the group to take action – the most important outcome of this process.

- Facilitate the group to monitor improvement in the welfare status of their animals regularly and to sustain welfare improvements through peer encouragement and peer pressure for continued action.

The Participatory Rural Appraisal (PRA) tools and exercises described in this section are all different but are designed to lead to a list of signs of good and poor welfare, created by the animal owners themselves, which can be observed directly (animal-based indicators). These signs are then used by the group to assess and monitor the welfare of all the working animals belonging to group members. Some tools also generate lists of the management practices, resources and services needed to produce welfare improvements: these are called resource-based indicators of welfare. The tools may also have other practical uses for the group, such as deciding what to look for when purchasing new animals, or how to increase the value of their working animals in order to sell them for a profit.

The diagram and table below explain the four steps of participatory welfare needs assessment in detail.

Step 3.1 Analysing how animals feel and what they need for their well-being

Step 3.2 Generating a list of animal-based and resource-based indicators. Agreeing on how they will be scored

Step 3.3 Observing animals and recording their welfare status

Step 3.4 Analysing the present welfare status of animals belonging to the group

Figure 4.3 Phase 3: Participatory animal welfare needs assessment

Phase 3 Participatory animal welfare needs assessment

Step 3.1

Analysing how animals feel and what they need for their well-being

Purpose
- To enable the group to build a common understanding of welfare based on animal needs and feelings.
- To enable the group to recognize how aspects of good welfare and poor welfare are expressed in an animal's appearance and behaviour.

Step 3.2

Generating a list of animal-based and resource-based indicators of welfare and agreeing on how they will be scored

Purpose
- To enable the group to summarize animal- and resource-based indicators in a format that enables assessment of animal welfare and monitoring of changes.

Step 3.3

Observing animals and recording their welfare status

Purpose
- To summarize all findings, giving a clear picture of the welfare of individual animals and of issues that affect all animals belonging to the group.
- To enable further reflection, discussion and decision-making.

Step 3.4

Analysing the present welfare status of animals belonging to the group

Purpose
- To enable the group to discuss and analyse the welfare issues of individual animals and the issues which affect all animals belonging to the group.

Process
- organize a group meeting;
- identify animal welfare needs, the effects which occur when these needs are not being met and how these effects can be seen on the animal through physical signs and behaviour (animal-based welfare indicators) ;
- transfer the animals' needs (resource-based welfare indicators) and animal-based welfare indicators to a list;
- reach a common agreement on how each indicator will be defined and scored;
- observe the animals and resources directly through a transect walk;
- record welfare indicators and their severity for each animal;
- discuss and collectively summarize individual and group welfare issues.

Tools
- Matrix ranking of animal welfare issues (T9)
- 'If I were a horse' (T17)
- How to increase the value of my animal (T18)
- Animal feelings analysis (T19)
- Animal body mapping (T20)
- Animal welfare practice gap analysis (T21)
- Animal welfare transect walk (T22)

Table 4.4 Process overview Phase 3: Participatory animal welfare needs assessment

Step 3.1: Analysing how animals feel and what they need for their well-being

Step 3.1 enables the group of animal owners to identify everything that their working animals need to live a better life. You can do this using one or more of the following tools from the Toolkit in Part III of this manual:

- 'If I were a horse' (T17);
- How to increase the value of my animal (T18);
- Animal body mapping (T20);
- Animal welfare practice gap analysis (T21).

These tools will help the group to:

- identify the needs of working animals;
- analyse how far the animals' needs are being met by their owners, users, carers and local service-providers;
- analyse the effects on working animals when their basic needs are not fulfilled;
- identify the physical and behavioural signs of each need, which can be observed directly, when an animal's needs are met or not met (animal-based indicators of welfare, Chapter 2).

We find that initially most owners only mention resources and services when considering the well-being of their animals. The presence of resources and services are important but they do not guarantee good welfare. For example, an animal may be given food (a resource), but if it is frightened of another animal nearby, or it has a sore mouth, it will not eat. These tools help to move people from looking at animal-related resources and services to observing the animal directly and seeing what it can tell them about its own welfare. They put the animal itself at the centre of analysis.

A detailed explanation of how to use the tools can be found in Part IV. Have a look at these and choose which ones would be the most suitable for the group you are working with.

On many occasions we start the process of discussing animal needs using the Animal body mapping tool (T20). The group draws the body of a working animal, identifies its body parts and then describes what is needed to keep each body part healthy.

The things animals need for their well-being include good management practices and kind behaviour of their owners, such as cleaning the animal shed, regular grooming, rest and calm handling. They also need good resources and services such as clean water, shelter and veterinary care. A fuller list will be brought out by the group during these exercises and can also be found in Theory box 3 on page 24.

Good Resources

Good handling and management

Figure 4.4 Animals need good resources and good handling and management for their well-being

As a facilitator, your aim is for the group to identify:

- the animals' needs;
- the effects on an animal when its needs are not met;
- where these effects can be seen on the animal's body or in its behaviour.

The illustration below shows some real results produced by a community group using the 'If I were a horse' tool (T17) for participatory welfare needs assessment. Here is an explanation of the illustration:

- Circle 1 (inner circle) shows what the animal needs or expects from its owner and other people – these are sometimes called resource-based indicators of welfare.

- Circle 2 shows the owners' present management practices, using stones to score how well each need is met.

- Circle 3 shows the effect on the animal when its needs are not being met.

- Circle 4 (the outer circle) describes signs that can be observed directly on the body or in the behaviour of the animal which would enable the group to know when each need is not being met – called animal-based indicators of welfare. In this example the participants identified that farriery needs to be done every 15 days. If it is not done the effect is lameness, which can be seen on the leg and hoof of the animal.

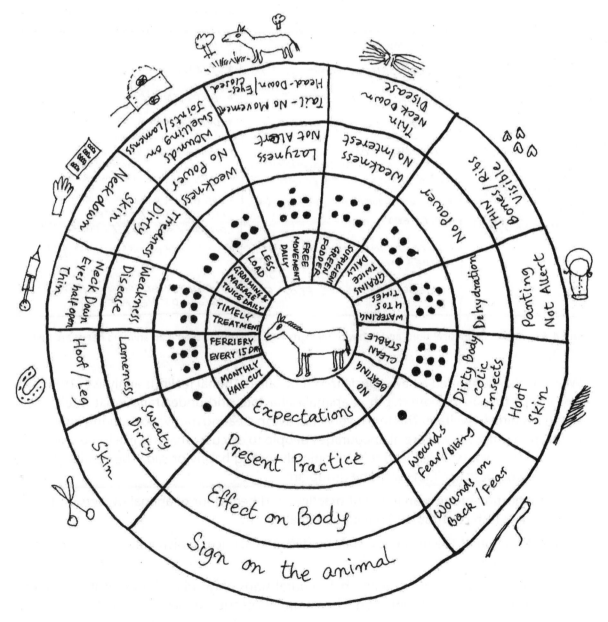

Figure 4.5 'If I were a horse' tool (T17)

Working out how animals feel – understanding the 'voice of the animal'

In Figure 4.5 above, the signs or welfare indicators visible on the animal (animal-based indicators of welfare) are of two different types:

1. Physical signs or symptoms, identified by looking at the animal's whole body, from head to tail.

2. Behavioural expressions or 'body language'. These can be seen in the body posture (how the animal stands) and in the position and movement of the eyes, ears, head, legs and tail. They can also be seen in the animal's behaviour and the way that it interacts with other animals or people. In Figure 4.5 these are: neck down, eyes half open, not alert and no movement of the tail.

Using these tools will generate many more indicators which vary from group to group.

Figure 4.6 below gives some examples of physical signs and behaviour which might be seen if an animal's needs are being met or are not being met.

	WHEN ANIMALS NEEDS <u>ARE</u> MET	WHEN ANIMALS NEEDS <u>ARE NOT</u> MET
EXAMPLES OF PHYSICAL SIGNS VISIBLE ON THE BODY	• NO WOUNDS • CLEAN, SHINY COAT • HARD, SMOOTH HOOVES • CLEAN, BRIGHT EYES	• WOUNDS • TICKS • BROKEN HOOVES • DIARRHOEA
EXAMPLES OF BEHAVIOURAL EXPRESSIONS OR 'BODY LANGUAGE'	• INTERESTED IN SURROUNDINGS • EARS MOSTLY UP OR FORWARDS • BEARING WEIGHT EVENLY ON ALL FEET • INTERACTING SOCIALLY WITH OTHER ANIMALS OF THE SAME TYPE • FRIENDLY TOWARDS PEOPLE	• NOT INTERESTED IN SURROUNDINGS • EARS MOSTLY DOWN OR BACK • HOLDING UP A LEG OR SHIFTING ITS WEIGHT BETWEEN DIFFERENT LEGS FREQUENTLY • STAYING ON ITS OWN A LOT • FRIGHTENED OR AGGRESSIVE TOWARDS PEOPLE

Figure 4.6 Examples of physical signs and behavioural signs which might be seen if an animal's needs are being met or not

We have also developed a new, specialized tool called Animal feelings analysis (T19) which helps groups to look in more depth at animal behaviour and understand what it tells them about their animals' welfare. This tool is very effective in creating motivation among individuals and groups to improve welfare. It encourages people to discuss what it is like to be a working animal and the factors which lead to their animals feeling happy or sad. The results of using one or more of these five tools with the group are:

• Identification of the management practices and behaviour of animal owners and other people which contribute to meeting animal needs.

• Identification of the resources, stakeholders and services which contribute to meeting animal needs.

• Identification of the physical and behavioural signs of good and poor welfare (animal-based indicators) which are seen when animals' needs are met or not met.

At this stage the group is ready for the next step in the process of Participatory welfare needs assessment.

Step 3.2: Generating a list of indicators of welfare and agreeing on how they will be scored

This step helps the group to arrange their animal-, management practice- and resource-based indicators into a practical format for assessing animal welfare and monitoring changes, as follows.

1. Participants write or draw the indicators identified in Phase 3, Step 3.1 (above) as a list which can be used for assessing their own animals.

2. The group then comes to a consensus on how each indicator will be scored and the exact definition for each score.

Using the results from the tools and exercises in Step 3.1, encourage the group to make a list of the parts of each animal that they will look at to assess its welfare, and the specific signs or animal-based indicators which they will look for. They may decide to look for weakness by checking for ribs showing and a low neck position, and to look at the legs for lameness and swelling.

They should then add management practices and resource-based indicators to the list, again specifying exactly what they will look for. For example, participants might want to find out how often the animal is offered water, examine the feeding trough for cleanliness, see how much feed is available in it and whether the height is right for the animal, and check the shed or stable for cleanliness, flies and smell (see Figure 4.7).

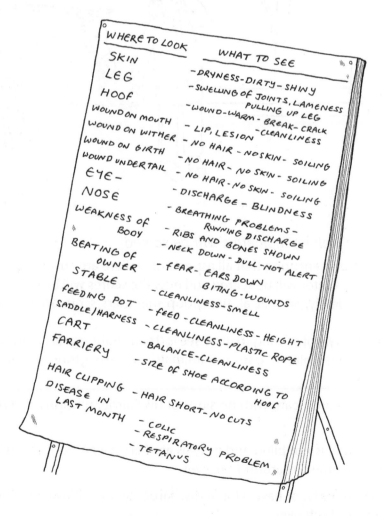

Figure 4.7 List of animal-based indicators, management practices and resource-based indicators of welfare

As the facilitator, you have an important role here: to check that the list represents all aspects of animal welfare. You can do this yourself by reviewing Chapter 2, especially the welfare frameworks described in Theory box 6 on page 41. After the group has finalized their list, sit together with them and check whether all aspects of welfare are covered. Sometimes we find that the owners' checklist only contains signs of physical welfare. If we notice this then the Animal feelings tool (T20) is very useful for bringing out signs of good and poor mental or emotional welfare more clearly.

Animal-owning groups can check their list of welfare issues is complete by putting them into categories:

- animal body, behaviour and feelings (including disease issues);

- management practices and behaviour of owners towards their animals;

- resources, stakeholders and services.

Agreeing on how the welfare indicators will be scored

After finishing the list and ensuring that all aspects of welfare are covered, the group needs to decide on how each indicator will be scored and how the scores will be defined. It is important to define each score clearly, so that all the group members score animals in the same way every time.

The most common type of scoring that we see is 'traffic light signals': red, yellow and green dots or marks. As an example, in assessing animals' eyes the group members might define traffic light scores as:

- Green: both eyes are clean, with no signs of discharge, no signs of dirt and the animal is not blind.

- Yellow: the animal is not blind and its eyes are fairly clean but there is a little discharge or watering, or some dirt around the eyes.

- Red: the eyes are dirty, or have a lot of discharge, or are whitish in colour, or the animal is blind in one or both eyes.

For assessment of the feet and hooves, scores might be defined as:

- Green: all four hooves are clean and well trimmed with no cracks or bad smell. If the animal is shod, the shoes are well set on.

- Yellow: one or more hooves are not well cleaned and trimmed, or have small cracks. If the animal is shod, the shoes are worn or need replacing.

- Red: one or more hooves are dirty, too long, have large cracks or a bad smell. If the animal is shod, the shoes are broken, missing or not set on properly.

Another method of scoring is to use numbers. The scale is defined and agreed by the group, who may decide that a score of 3 means bad condition, 2 means medium condition and 1 means good condition. Since illiteracy is common, animal owners may find this type of scoring more difficult than the traffic lights, although they are often comfortable with recognizing and writing numbers or tally marks. We have also seen groups choosing a combination of numbers and signs which they find easy to record, such as scoring 2 for poor condition, 1 for medium condition and a plus sign (+) for good condition. Groups who use numerical scoring usually make a summary score for each animal and for the village as a whole. This has an added value for analysis (Phase 3 Step 3.4).

In our experience, most groups start with the traffic light scoring system and some then change to numerical scoring. The shift to a different scoring system happens as the group gradually builds up experience in assessing their animals' welfare and recording the information. These groups find that the traffic lights are not sensitive enough to make a good assessment of the severity of welfare issues in their animals, so they develop more in-depth scoring. Several of our groups are now scoring welfare issues on a scale of 1-5 or 1-10 and defining the parameters for each score in great detail. In many cases the groups started to use this more accurate welfare scoring in competitions to select the best animals belonging to their group or village.

Step 3.3: Observing animals and recording their welfare status through an 'animal welfare transect walk'

Step 3.3 enables participants to assess the real welfare issues of individual animals belonging to group members, and also to identify any common issues which affect all animals belonging to the group.

Once all the animal-based and resource-based indicators are listed, go with the group on an Animal welfare transect walk (T22), by walking from house to house through their village. Encourage the members to check all the animals belonging to the group, one by one. The group assesses each animal together and agrees the score for each indicator on their list. Scores are recorded on a chart or register which is kept by the group for regular monitoring and follow up. Figure 4.8 shows the traffic light scoring chart resulting from issues identified using the 'If I were a horse' tool (T17) and the list of indicators described in Steps 3.1 and 3.2.

The Animal welfare transect walk (T22) may be done by men, women or both together, according to the preference of the group and based on the findings of other tools, such as the Daily activity schedule (T4). In some countries, women do most of the animal-related tasks around the home and men manage the animals while working. We have found that groups of women are particularly meticulous, serious and effective in carrying out the Animal welfare transect walk (T22) exercise collectively for their animals.

WHERE TO LOOK	WHAT TO SEE 1-16 JAN	NAMES OF OWNERS O GREEN ◎ ORANGE ● RED								
		Guddu	Harpal Singh	Taipal Singh	Amichand	Satyapal Singh	Gopal Singh	Jeevanlal	Vijay Singh	Hemraj
SKIN	DRYNESS-DIRTY	◎	●	O	O	O	◎	O	O	O
LEG	SWELLING OF JOINTS	O	O	O	O	O	O	O	O	O
HOOF	LAMENESS-PULLING UP LEG	O	O	O	●	O	O	O	O	O
HOOF	WOUND-WARM	O	O	◎	●	O	O	O	O	O
HOOF	BREAK-CRACK	O	O	O	O	O	●	O	●	O
HOOF	CLEANLINESS	◎	◎	◎	●	●	O	O	●	●
WOUND ON MOUTH	LIP-NO SKIN	O	O	O	O	O	●	O	O	O
WOUND ON GIRTH	NO HAIR-NO SKIN	O	O	O	O	O	●	O	O	O
WOUND ON WITHER	NO HAIR-NO SKIN	O	O	O	O	O	O	O	O	O
WOUND UNDER TAIL	NO HAIR-NO SKIN	O	O	O	O	◎	O	O	O	O
EYE	REDNESS-DISCHARGE BLINDNESS	O	O	O	O	O	◎	◎	●	◎
NOSE	BREATHING PROBLEM RUNNING DISCHARGE	◎	O	◎	O	O	O	●	●	◎
WEAKNESS OF BODY	RIBS AND BONES SHOWN	◎	◎	◎	O	O	◎	O	◎	O
NECK	DOWN-DULL ALERTNESS	O	O	O	O	O	O	O	O	O
BEATING	EARS DOWN-FEAR BITING-WOUNDS	O	O	O	O	O	O	O	O	O
STABLE	CLEANLINESS-SMELL	O	◎	◎	◎	◎	O	●	O	O
FEEDING POT	FEED-CLEANLINESS HEIGHT	◎	◎	◎	●	◎	◎	◎	◎	◎
SADDLE/HARNESS	CLEANLINESS-SMELL-NOT SHARP	O	◎	O	O	O	◎	O	O	O
CART	BALANCE-CLEANLINESS	O	◎	●	O	O	◎	O	◎	●
FARRIERY	SIZE OF SHOE ACCORDING TO HOOF	O	O	O	O	O	O	O	O	O
HAIR CLIPPING	HAIR SHORT NO BRETING EA	●	O	O	●	O	O	O	●	O
DISEASE IN LAST ONE MONTH										
	COLIC	O	O	O	O	O	O	O	O	O
	RESPIRATORY PROBLEM	●	●	●	●	●	●	●	●	●
	TETANUS	O	O	O	O	O	O	O	O	O

Figure 4.8 Animal welfare transect walk (T22) recording chart

Step 3.4: Analysing the present welfare status of animals belonging to the group

The purpose of Step 3.4, the last in Phase 3, is to summarize all the findings, giving a clear picture of the welfare of individual animals and of the issues affecting all the animals belonging to the group. This enables further reflection, discussion and decision-making on individual and collective action to improve welfare.

As soon as the Animal welfare transect walk (T22) is complete, the group sits together and summarizes the findings on their chart to draw conclusions. If the walk has been carried out on more than one day, it is useful to hold a group discussion at the end of each day, with a final meeting on the last day of the exercise. The group summarizes the findings for each individual animal and for all the animals together. In particular, the group draws out the welfare issues which are scored red or bad condition, for individual animals and for the village as a whole. This will generate a list of welfare issues which will be used in Phase 4.

The process of joint analysis leads to individual as well as collective action, because awareness of animal welfare issues is awakened and peer pressure for action is generated by and among group members. Both of these drive the group towards appropriate planning for intervention.

The advantage of this intensive approach is that, in most cases, action to improve the welfare of working animals begins immediately after completing the Animal welfare transect walk. This is because individuals realize the welfare issues facing their animals and are aware of the atmosphere of positive competition generated between group members.

Phase 4. Community action planning

The purpose of Phase 4 is to move with the group from their new awareness of animal welfare issues, identified through the exercises carried out so far, towards individual and collective action for improvement.

There are three steps involved in this process:

4.1 Prioritizing welfare issues of importance to working animals and their owners

4.2 Root cause analysis of welfare issues

4.3 Preparing a collective plan of action to improve the issues, based on the root cause analysis.

Phase 4 Community action planning

Step 4.1 Prioritizing welfare issues of importance to working animals and their owners

Purpose
- To list all the animal welfare issues identified in the previous phases
- To prioritize these issues according to importance

Step 4.2 Root cause analysis of welfare issues

Purpose
- To analyse the causes or contributing factors to major welfare issues, in order to determine which factors need action or intervention

Step 4.3 Preparing a collective plan of action to improve the welfare issues

Purpose
- To identify the action needed to tackle each root cause.
- To plan:
  o the actions to be taken against each root cause
  o who will carry out the actions
  o when to carry out the actions
  o who will monitor that the action is taken as agreed
- To develop measuring indicators for each activity to be monitored by the group.

Process
- Organize a group meeting with animal owners, carers and other stakeholders as appropriate.
- Prepare a list of welfare issues, based on the Animal Welfare Transect Walk recording chart described in Phase 3, and any other issues identified and discussed during previous exercises.
- Prioritize issues on the basis of how common they are, how severely they affect working animals and which need immediate, medium-term and long-term solutions
- Identify the root causes which are responsible for each issue.
- Collectively look for possible solutions to the major root causes of the welfare issues (express these solutions as actions which can be taken by the group, individually or together).
- Discuss the roles and responsibilities of each group member in carrying out each action.
- Agree the time frame for action, the resources needed, requirements for support from group members and any external support needed
- Agree how to measure and who will monitor the implementation of the action plan
- Present the group action plan to the wider community, along with a formal or informal agreement to carry out the action plan.

Tools
- Pair-wise ranking (T8)
- Matrix ranking (T9)
- Three pile sorting (T23)
- Animal welfare story with a gap (T24)
- Problem horse (T25)
- Animal welfare cause and effect diagram (T26)

Table 4.5 Process overview Phase 4: Community action planning

Step 4.1: Prioritizing welfare issues of importance to working animals and their owners

Welfare issues affecting their working animals have now been identified by the group. Their next task is to decide which problems are the most serious or important to change, by ranking issues according to their importance. This helps the group members to set a realistic agenda for their own actions within limited financial and other resources.

During analysis of the charts recorded during the Animal welfare transect walk (T22; see Phase 3), the group produced a list of welfare issues which scored red (or bad welfare) on their scoring system. Encourage the participants to reflect back on the other exercises they have carried out together. Add the animal-related problems identified in those exercises to the list as well.

You can facilitate the group to rank these welfare issues according to their priorities for taking action, by using simple discussion or by writing the issues on separate cards and agreeing an order of preference (also known as Preference ranking). Alternatively you may use Pair-wise ranking (T8) or Matrix ranking (T9). Pair-wise ranking enables people to decide on priority issues by comparing each issue against the others. Sometimes groups may find this difficult and be unable to come to a consensus, in which case we find that Matrix ranking works better. In our experience many groups use a combination of tools: Preference ranking followed by Matrix ranking.

It is important that issues are prioritized according to the preferences of the animal owners, rather than according to your priorities or those of your supporting agency. If the group does not have ownership of decisions, action will not follow.

The list of animal welfare issues is often a long one and the group cannot act on all of them at the same time. It may be useful to sort the list into problems which need immediate action (within one month) and those which are medium-term goals (action to be taken within about one year) or long-term goals (two years or more).

Process box 5. What do we do about welfare issues that are incurable?

We discovered that during analysis of the traffic light charts recorded during Animal welfare transect walks (T22), groups often separate out animal welfare issues which are marked red but are not curable, such as blindness. Although blindness is not reversible, it is a welfare issue that needs to be acted upon and should not be set aside. Our approach is to trigger a discussion on why this particular animal became blind, and any other causes for blindness in working animals. This leads to action to prevent other animals from losing their sight in the future. They also encourage production of a welfare action plan for the affected animal, including how best to handle a blind animal during work, such as training it to respond easily to voice commands and using streets with less traffic. The plan also covers how best to manage the blind animal at home and during rest periods, for example by putting its food and water in the same place every day and ensuring that it is not chased away from food by other animals.

Step 4.2: Root cause analysis

The next step for you and the group is to identify the underlying causes for their priority welfare issues. This includes factors contributing to poor welfare while animals are working and also during their rest periods. A variety of tools may be helpful for this purpose; the two that we commonly use are Problem horse (T25) and Animal welfare cause and effect diagram (T26). These tools establish a hierarchy of root causes for poor working animal welfare and effects of poor welfare, constructing a 'tree' which shows the cause-effect relationships. It is very helpful to consider the effects of contributing factors on the owner and his or her family, as well as on the animal. This increases people's motivation to take action for change. For example, discussing the causes of wounds on specific parts of a working animal's body may highlight causal factors such as the structure of the harness and how clean it is, the size of the yoke or saddle tree, or the design of the cart. Effects of these factors on the animal may include pain, weight loss and reduced working capacity. Effects of the animal's wounds on its owner could include less income (from reduced work and increased expenditure on medicines) or lower status in the community.

In the early stages, most causes identified by the group are likely to be superficial. In the case of wounds on the back or chest of the animal, participants may perceive them as easy to deal with and the action they agree might be regular cleaning of the wounds. However, in working animals, cleaning of these wounds is not usually enough to cure them. They may have multiple root causes such as ill-fitting harness, dirty padding, an unbalanced cart, rough road conditions, over-loading or beating, and it is not likely that these causes will come out in the initial discussion. The group has to experience for itself that despite all efforts to clean the wounds, some do not get better. This experience triggers them to carry out a more in-depth analysis of the causes of wounds. It is important that you are patient and encourage the group to reflect carefully and to analyse the causes or contributing factors in detail. If the discussion is pushed too fast, people may not have time to come out with the real underlying causes (see Case study H on page 106).

Step 4.3: Preparing a collective plan of action to improve welfare issues

Once the group has agreed on the root causes for each key problem, Step 4.3 is to make a community action plan for implementing change, creating an action for each root cause identified. The community action plan serves as an open community 'contract' for action which spells out the steps which will bring about change. This notes who will take which action and when, helping people to take responsibility for addressing problems in a systematic way and building in accountability.

It also brings agreement on the support that they need in order to implement the plan: from you, your supporting organization and other external institutions or stakeholders. At this stage, encourage the group to discuss past experiences or previous efforts made to tackle issues, so that their lessons learned can contribute to the new plan.

The major contributors to the community action plan should be the animal owners' group, because the plan is designed to reflect their interests. However, it is always helpful to include representatives of local stakeholders or service providers, such as the farrier, health provider, women's group and other concerned agencies. It is particularly important to involve women and children at every stage of the planning process, either together with the men or separately. This not only ensures better results, it also leads to sustainability in the long term because they will often be the ones taking care of animals in their non-working hours.

Sometimes a simple plan may be prepared by the group, based on one or two very urgent issues for immediate action, or those of particular interest to the members. When these issues are sorted out they will start to tackle the next ones on their priority list, and so on. As the group becomes stronger and more confident they may wish to produce a more comprehensive action plan.

Whatever the size of the plan, it must be specific if it is to be a useful guide for change. This includes:

- the welfare issue identified

- the cause(s) of each issue (based on the root cause analysis)

- action to be taken against each root cause

- who will do them (clear roles and responsibilities)

- when to do them (time frame)

- who will monitor that the action is really taken as agreed

Depending upon the group size and dynamics, everyone may wish to work together or to break into smaller sub-groups. These can work on action to overcome different root causes, or on different parts of the plan, according to their choice. Sub-groups come back together once they have completed their section, to comment on each others' work and offer suggestions for changes. Check with them that the animal body, behaviour and feelings issues, the management and owner behaviour issues and the resource, stakeholder and service issues prioritized in Step 4.1 are all covered.

ISSUE	CAUSE	ACTION TO BE TAKEN	WHO?	WHEN?	WHO WILL MONITOR?
DIRTY SKIN					
CLEANLINESS OF HOOF (MAGGOTS)					
DIRTY EYES					
WOUND UNDER STOMACH (GIRTH)					
BALANCE OF CART					
RESPIRATORY PROBLEM					

NAME	ISSUE	CAUSE	ACTIONS TO BE TAKEN	WHO?	WHEN?	WHO WILL MONITOR?
Guddu						
Jaipal						
Harpal						
Amichand						
Hemraj						

Figure 4.9 Two possible formats for a community action plan

We have experienced that this action planning leads two types of action:

Individual action by each member of the group to benefit their own animal, such as cleaning of the saddle and harness, repairing and balancing the cart and cleaning any wounds on the animal. These individual actions are decided collectively by the group and are monitored by the group.

Collective action by the whole group to benefit all the animals belonging to group members, such as organizing vaccination of all animals at the same time, or building a water trough at their work congregation point.

At this stage it is vital to discuss how implementation of the action plan will be measured and monitored. Monitoring is most effective when group members agree to monitor each other. This brings in peer pressure and peer encouragement for action.

- What measures will be used to show that people are doing what they have agreed to do (activity monitoring)?

- Who will measure each activity? This involves other community members in supporting the action and ensuring that it is accomplished.

- When will they be measured? The frequency of monitoring depends on the group's preference – they could monitor each other weekly, fortnightly or monthly.

COMMUNITY LED TETANUS VACCINATION				
NAME GROUP MEMBER	Date 1st Vaccination	Date 2nd Vaccination	1st Booster	2nd Booster
Harpal	19.8.08	17.9.08	18.9.09	
Jaipal	19.8.08	17.9.08	18.9.09	
Raipal	19.8.08	17.9.08	18.9.09	
Amichand	19.8.08	17.9.08	18.9.09	
Jeevanlal	19.8.08	17.9.08	18.9.09	
Satyapal	17.9.08	18.10.08	18.10.09	
Gopal	19.8.08	17.9.08	18.9.09	
Brijendra	19.8.08	17.9.08	18.9.09	
Hemraj	17.9.08	18.10.08	18.10.09	

The full group then agrees a final community action plan. See Figure 4.10 below.

ISSUE	CAUSE	ACTION TO BE TAKEN	WHO?	WHEN?	WHO WILL MONITOR?
DIRTY SKIN	• Dirty stable • No regular grooming • No hair cut • No bathing	Daily cleaning of stable Grooming every day after work Bathing once in a week	Harpal Singh Gopal Singh	From tomorrow onwards	Hemraj, Jaipal
CLEANLINESS OF HOOF (MAGGOTS)	• Dirty stable • No regular cleaning of hoof • No regular checking of hoof	Proper cleaning of stable every day Application of lime Regular cleaning of hoof using hoof cleaner	Guddu, Harpal, Jaipal, Amichand, Satyapal, Vijay, Hemraj	Immediately	Jeevanlal, Gopal
DIRTY EYES	• Working in dusty environment • No cleaning	Cleaning eyes twice a day	Jeevanlal, Gopal, Vijay, Hemraj	During brick kiln work	Guddu, Harpal
WOUND UNDER STOMACH (GIRTH)	• Use of hard belt • Belt not cleaned, dirty • No oiling of belt	Replace the strap belt (rope) with soft cloth strap or leather Daily cleaning of belt Regular oiling of belt	Gopal	Within days, check every day if wound reduces	Vijay, Amichand
BALANCE OF CART	• Tyre pressure too low and different for each tyre • Uneven load	Check pressure every month Same pressure at both tyres Divide load evenly when loading	Harpal, Jaipal, Gopal, Vijay, Hemraj	From tomorrow onwards	Guddu, Harpal
RESPIRATORY PROBLEM	• Not known • Dirty stable	Invite vet to come to meeting to discuss prevention	All	In February– Sunday 5th	Satyapal, Jaipal

Figure 4.10 Example of a community action plan

Case study H. Surprises in Khanjarpur village

Source: Ramesh Ranjan, Brooke India, Ghaziabad, Uttar Pradesh, December 2007

Brooke India's Ghaziabad District Equine Welfare Unit started to work in Khanjarpur village in 2007. At a cart-horse owners' meeting in July, group members started to discuss who among the eleven cart owners was a good owner and who was the best owner in the whole village. They agreed to set criteria on which to base their decision, including whose animal had a clean stable, fewest wounds, good condition of the hair, a saddle in good condition, a balanced cart with good tyre pressures, an owner who didn't beat their animal, and many others. It was decided to score each owner according to these criteria, by going on an Animal welfare transect walk to each person's doorstep to see the animal and its stable. With all of the owners walking together, this was an exciting exercise. Findings were shared and there was another long discussion. The maximum score they gave was 8 out of 10 for Mr. Jai Prakash's horse, as it had no wounds at all.

Wounds then became the main topic of discussion, since all the animals were suffering badly from them except for Jai Prakash's horse. People were keen to find out how to improve the wound condition of their animals. This led them to search for factors causing the most severe wounds and to decide on specific actions, such as cleaning and softening the *Kathi* (saddle tree), tying clean cloth padding around it, oiling the leather part of the *Kathi* regularly, giving more water to the animals, balancing carts properly and maintaining appropriate tyre pressures.

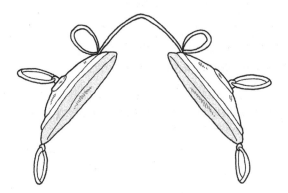

When the group repeated the transect walk in August, most of the animals' wounds had decreased in size and severity, but the wounds on the withers of Brijpal's and Rakesh's horses had increased. Both owners were adamant that they had been taking the agreed care of their animals, harnesses and carts. This compelled the group to investigate further. They concluded that the shape and size of the *bangla* (part of the saddle tree) also influenced wounds and that they had missed this in their earlier discussions. The *bangla* on every saddle tree was measured and matched with the size of the animal. Everybody was curious to know the outcome. To everyone's surprise, the size of the *bangla* on Brijpal's and Rakesh's saddle was almost double that of the others. Both owners accepted the findings cautiously and each changed their saddle tree for one with the right size of *bangla* in relation to the size of their horse. The Animal welfare transect walk was repeated for a third time in October, and this time the wounds on Rakesh's and Brijpal's horses had reduced and almost all of the other horses had no wounds at all. The animal owners had found the right solution to their problem.

This experience really helped the Khanjarpur horse owners' group to become confident and built their capacity to solve their own problems. They continue the practice of analysing issues jointly, so their success not only changed the wound situation but also improved many other management practices for the benefit of their working animals.

Phase 5. Action and reflection

The purpose of Phase 5 is to help the group to implement their community action plan, monitor it regularly and reflect on their findings and experiences together.

It is essential for the group to critically appraise the performance of both the individual members and the group as a whole, in order for the animal welfare intervention to succeed. These positive, constructive appraisals translate action into learning which in turn translates into further action. The depth of reflection has a major effect on the quality of the action that follows.

Periodic tracking of progress can help animal owners and carers to:

- build their interest in the intervention and their commitment to making it work

- assess the roles of different stakeholders

- understand the changing dynamics in their environment

- generate increasing knowledge about actions which work or don't work in their community action plan, leading to corrective action or improvement

- share responsibility for dealing with challenges

- bring peer pressure and peer motivation to influence individual actions

- trigger greater understanding, sensitivity and care for their working animals

This will also enable you and your supporting organization to understand their situations and constraints more clearly. Two types of collective monitoring are essential for the success of the action and reflection process:

Monitoring group activities
Monitoring of group activities is a regular function of every group meeting from Phase 5 onwards. This checks that group members and other stakeholders are doing what they agreed to do in their action plan.

Monitoring changes in the welfare of working animals
Monitoring of changes in animal welfare as a result of the activities carried out is achieved by repeating the Animal welfare transect walk exercise (T22).

In the following illustration and table the process of action and reflection is explained in more detail.

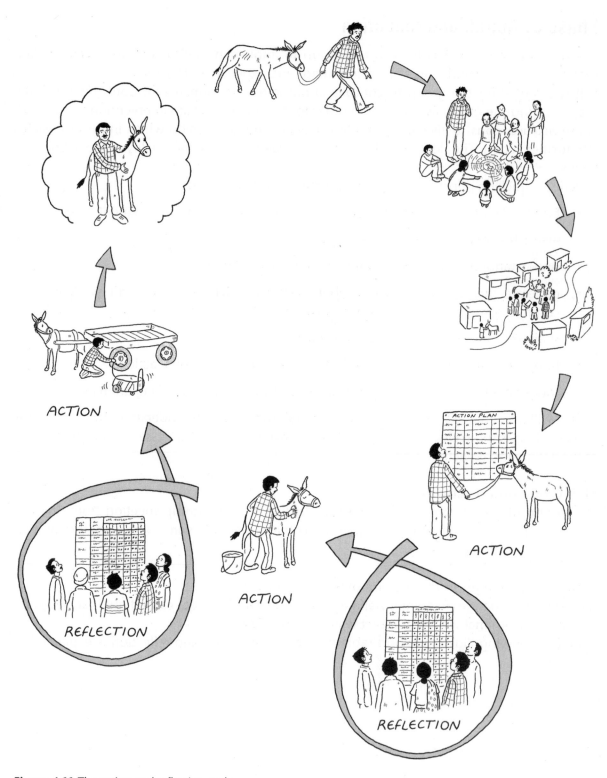

Figure 4.11 The action and reflection cycle

Phase 5 Action and reflection

Step 5.1 Implementation and monitoring of activities in the community action plan

Purpose
- To enable the group to take individual and collective action on the activities agreed in the community action plan
- To check that activities are carried out, through community and group meetings and home or site visits

Process
- Organize regular group meetings to review individual and collective efforts towards welfare improvement.
- Check and record the activities agreed in the community action plan to ensure that they are carried out.
- Generate resources needed for implementation of the plan, through collective contribution and by forming links with other resource providers.
- Initiate support for implementation of those activities which need external support.

Tools
- Pair-wise ranking (T8)
- Matrix scoring (T9)
- Animal welfare transect walk (T22)
- Problem horse (T25)
- Animal welfare cause and effect diagram (T26)

Step 5.2 Participatory monitoring of animal welfare changes, creating a cycle of reflection and action

Purpose
- To monitor the effect of the action plan on the welfare of working animals
- To reflect on the changes seen on the animal and develop new action points

Process
- Repeat the Animal welfare transect walk (T22) after one to three months, in the same way as the first time (Phase 3, Step 3.3).
- Analyse the results of the Animal welfare transect walk recording chart
- Take corrective action to keep the plan on track and/ or to develop new action points.

Table 4.6 Process overview Phase 5: Action and reflection

> ### Process box 6. A good action and reflection process
>
> - draws on local resources and capacities;
> - recognizes the innate wisdom and knowledge of the community;
> - demonstrates that animal owners are creative and knowledgeable about their animals and their situation;
> - ensures that other animal-related stakeholders are part of the decision-making process;
> - has facilitators who act as catalysts and who assist the community in their action and reflection.
>
> Capeling-Alakija, S., Lopes, C., Benbouali, A., Diallo, D. (1997) *A Participatory Evaluation Handbook – Who are the Question Makers?* OESP Handbook Series, Office of Evaluation and Strategic Planning, United Nations Development Programme, New York, USA.

Step 5.1: Implementation and monitoring of activities in the community action plan

This first step in the action-reflection process enables the group to take action on the activities agreed in their community action plan. A well drawn up community action plan leads to immediate action.

Your role as the facilitator is to contribute to the creation of an enthusiastic atmosphere and enabling environment where the group members can help each other with their agreed actions. This includes:

- Regular meetings to review individual and collective activities against the community action plan

- Generation of the resources needed to support their actions, for example through regular contribution of money to a common fund, or by creating links with other agencies, resource providers and government support schemes

- Generation of other external forms of support to implement their activities, if needed

- Maintenance of a record or register by the group, in which they record all of their decisions

At the beginning you may need to initiate this process of recording and gradually hand it over to group representatives. Where none of the group members are literate, they may decide to ask for help from a literate person or from school children in the village.

Step 5.2: Participatory monitoring of animal welfare changes, creating a cycle of reflection and action

Repeating the Animal welfare transect walk (T22; Phase 3, Step 3.3), at intervals of one, two or three months, enables the group to monitor changes in the welfare of their animals. Their scores for each animal welfare issue are recorded on the same monitoring chart each time (see Figure 4.12 below).

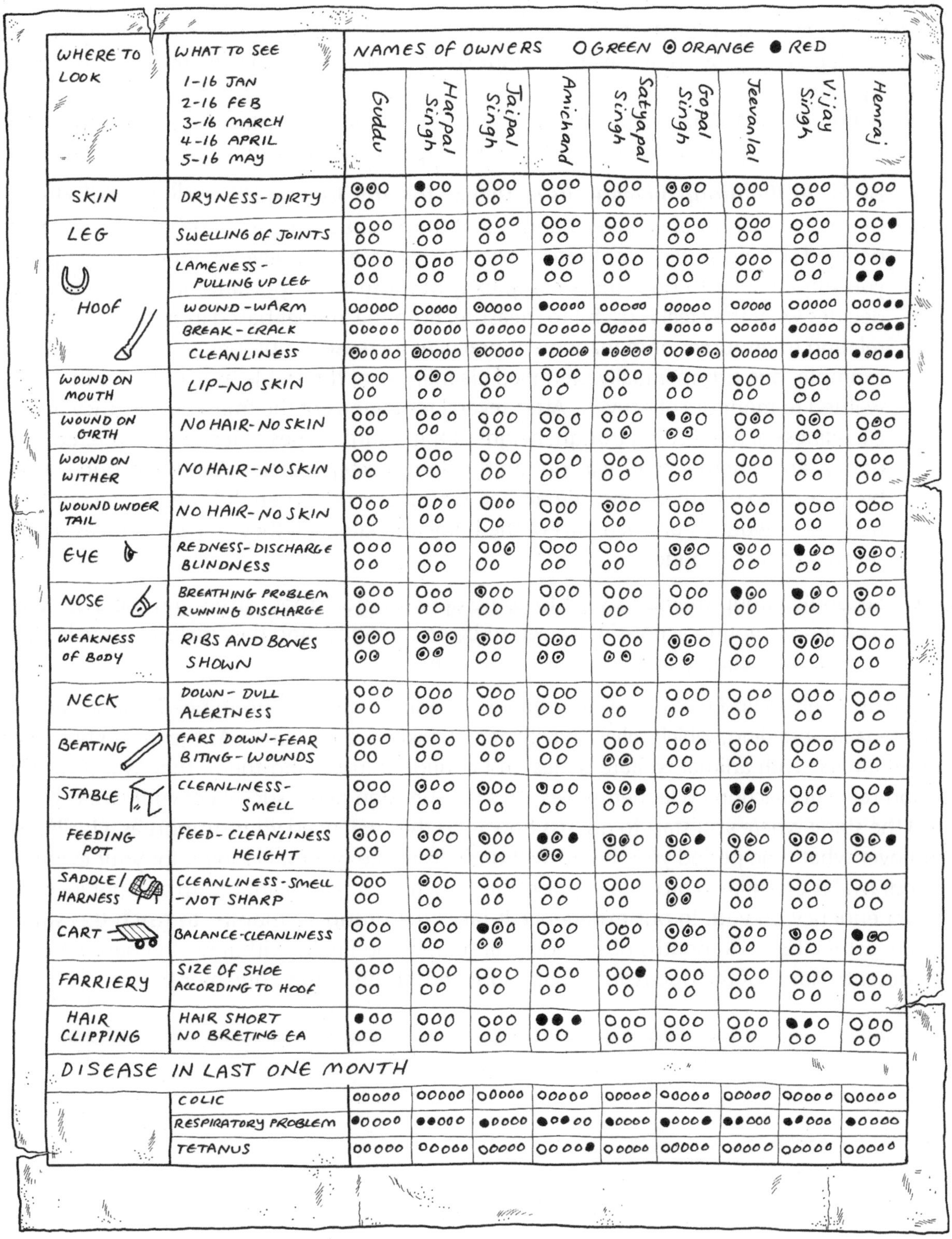

Figure 4.12 Recording chart showing the results of repeated animal welfare transect walks

Group members then sit together again to reflect on their findings, both positive and negative. Improvement in scores shows the effect of the actions they have taken to improve management of their working animals and to prevent welfare problems from occurring. They may identify gaps in their current practices, decide if further actions or closer monitoring are needed and record these.

In the communities where we work, we have found that many welfare issues will improve, but some will not change despite the group's action. This stimulates further discussion and an in-depth root cause analysis on these specific issues, using the Problem horse tool (T25) or Animal welfare cause and effect diagram (T26). This second level of root cause analysis is an essential step in the process of solving the more difficult or long-term welfare problems facing working animals.

Repetition of the Animal welfare transect walk (T22) leads to continuing refinement of the community action plan. We have found that two types of refinement commonly occur:

- Firstly, as the group increases its sensitivity towards its animals, the members choose to use a longer list of welfare changes that they want to measure and they create more detailed scoring systems for these. They will do these themselves in time. If they do not, you should not introduce more complexity because it is important that the community decides what they feel is useful to measure. Both the animal-based indicators and those relating to resources or management practices increase in number and complexity. For example, one group initially identified cleanliness of the stable as something that they wanted to monitor collectively. After several repetitions of the transect walk they added cleanliness of the feeding trough, height of the trough, types of feed being offered (wet or dry) and whether chalk was put on the floor to keep it clean. Another group started by measuring whether their animals had wounds or not. Then they began to count the number of wounds on each part of the body and later still they started to measure the severity of each wound.

- Secondly, group members start to come up with more root causes and the associated welfare-promoting actions that need to be encouraged, and they include these in their community action plan and recording system.

You are likely to find that in the early stages the animal-owning group needs lot of support and capacity-building, which relies heavily on your skill as a facilitator. As the group becomes more familiar with animal welfare issues and confident in solving them, they will drive this action-reflection-action cycle themselves. This is a sure sign of the success of your work. This is also the stage where you start to discuss how long they will need your support as a facilitator and over what time period you should withdraw from the group. Planning for your eventual withdrawal is essential in order to support the growth of a self-reliant group and not increase its dependency on you. In our experience it takes the group 12 months to reach this stage, and a further 12 to 18 months of strengthening until you finally withdraw. In Phase 6 below we describe some of the discussions that might contribute to this agreement.

Phase 6. Self-evaluation and gradual withdrawal of regular support

Phase 6 is about assessing the longer term impact of the group's efforts to improve the welfare of their working animals. This may be carried out at the mid-term (usually twelve to eighteen months) and at the end-point (usually two to three years) of your involvement with the group. It enables group members to see positive changing trends in animal welfare and reflect on any issues which may need further action with your continuing support. Your job is to check that the group can stand on its own feet before you withdraw.

During this phase you should come to an agreement with the group about how much of your support that they will need in the future. They will be working towards having much less input from you and longer intervals between your visits. Long-term support may include holding an annual meeting, helping to overcome specific problems or crises, linking the group with other relevant agencies and federating local community groups (see Chapter 5).

Gradually withdrawing your support and that of your organization will enable you to extend your facilitation into other villages or communities where working animals are in need. In the long term your projects will cover more animals over a larger area than would be possible if you stayed closely involved with one group or community.

Phase 6. Self-evaluation and gradual withdrawal of regular support

Step 6.1 Self-evaluation

Purpose

- To analyse the visible impact of group activities on animal welfare, the successes and challenges experienced during implementation of the community action plan, and the impact of animal welfare improvements on the livelihoods of the group.

Process

- Plan a meeting in advance, including all the stakeholders and service providers who have been part of the Action Plan
- Share successes, challenges and lessons learned from implementation of the community action plan
- Analyse the repeated Animal welfare transect walk chart to see the changes in working animal welfare
- Identify how the actions which owners took for the benefit of their animals improved their own lives
- Prepare an action plan to continue welfare improvement based on the self-evaluation analysis.

Tools

- Historical timeline (T7)
- Changing trends analysis (T11)
- Before-and-after analysis (T11)
- Group inter-loaning analysis (T14)
- Success and failure stories (see text)
- Most significant change technique (see Theory box 9)

Step 6.2 Gradual withdrawal of regular support

Purpose

- To develop a plan for gradual withdrawal of your regular support to the community.

Process

- Discuss the process of continuing improvement in working animal welfare with the group when they develop their new action plan in the previous step.
- Agree what support is needed from you and your supporting organization to implement their action plan.
- Agree a time frame for giving this support and implementing the plan.
- Establish criteria with the group for measuring their self-reliance and enable them to identify their current level of self-reliance based on these criteria.
- Initiate the formation of group clusters if possible.
- Withdraw your regular facilitation from the group according to the agreed time frame. Provide active support only in response to the group members' request and only in a crisis situation which they cannot resolve on their own.

Table 4.7 Process overview Phase 6: Self-evaluation and gradual withdrawal

Phase 6 consists of the two final steps 6.1 and 6.2.

Self-evaluation. The group stands back and looks at its progress over a longer period of time than it normally does at regular meetings. The group uses the Animal welfare transect walk recording chart to look for changing trends in the welfare of their working animals. They look at their successes and failures and the group's achievements and challenges during implementation of their community action plan. They also consider how the impact of animal welfare changes affects the livelihoods of the group. Then they agree on what actions to take to make these changes long-lasting.

Transition in the facilitator's role. You agree with the group on a transition from your regular facilitation at their meetings to a situation where they continue to meet and take action to improve the welfare of their working animals without your regular support.

Step 6.1: Self-evaluation

The main purpose of this step is to analyse the Animal welfare transect walk recording chart for evidence of changing trends in animal welfare together with the visible impact of animal welfare improvement on the lives of group members and their families and to analyse the successes and failures of the community action plan in achieving these changes.

Sharing of lessons learned leads to revision of the community action plan for future work. This is a self-evaluation process by the group although, unlike the short action-reflection cycle between group meetings (described in Phase 5), its aim is to evaluate progress broadly over longer periods of 12 months or more. During this process, the group makes changes to their community action plan which will help them to maintain welfare improvement in the long-term.

You will need to plan this meeting in advance because it will take longer than a regular group meeting. Some groups decide to hold a two-day meeting, while others plan to spend two hours every day for three to four days on the self-evaluation process. It is very useful to involve local stakeholders and service providers, such as veterinary workers, medicine shop owners, farriers, harness- and cart-makers and any others identified during Phase 2. Their involvement will help to strengthen the community action plan by encouraging them to continue working closely with the group on improving service provision for working animals.

During this meeting, three main areas will be evaluated:

- Success and failures of the community action plan.
- Changing trends in the welfare of their animals.
- The impact of improved animal welfare on the lives of animal owners, their families and the community.

Start the meeting by asking the group to remember what happened right at the beginning when their interventions started. Compile a Historical timeline (T7) of the events and challenges which occurred throughout the period since they began to work together. This will set the climate for in-depth discussion.

Successes and failures of the community action plan

In Phase 5 the group looked together at their activities, to find out whether they were carried out as agreed and whether they led to the desired change in a welfare issue. Now, in the Phase 6 evaluation meeting, the group looks back further, comparing the situation before they started to implement their community action plan with the situation now. This is easily done using the Changing trend analysis (T11). The group analyses which actions have been most effective and which less effective and the reasons why. This helps participants to learn, to change their interventions if necessary and to plan for continuing action. Matrix ranking (T9) can be used to compare the relative success of the activities taken up. During this part of the analysis it is also useful to discuss the achievements and difficulties faced while working together as a group, with reasons for these, and how any difficulties were overcome.

Changing trends in the welfare of their animals

The process that you have facilitated so far will have enabled owners to recognize when animal welfare is getting worse and to take action quickly, either as an individual or collectively. In order to create a deeper understanding, analysis of the Animal welfare transect walk (T22) monitoring charts will highlight the dynamic status and changing trends in the welfare of working animals belonging to the group. Members may notice changes in welfare according to the season, the animals' workload and other factors in their living and working conditions and their environment. Facilitating owners to recognize these changes and the effect of their own actions will lead to sustained improvement in the welfare of their animals.

Figure 4.13 below shows information compiled by an equine welfare self-help group from their Animal welfare transect walk (T22) monitoring charts. It contains the results of five transect walks carried out monthly between January and May 2009. The group looked at the change in number of welfare issues per individual animal owner over the five-month period (Figure 4.13a) and the change in the total number of welfare issues affecting all animals in the village (Figure 4.13b). This analysis enabled the group to identify and discuss persistent welfare problems which had not come out during their routine intervention monitoring (as described in Step 5) and as a result they planned some new actions. For example, the chart on the right shows that weakness of the body remains an issue with several animals in the village. This led to collective action to improve animal feeding (see T27, Analysis of animal feeding practices).

In some villages we have seen that this compilation and analysis of monitoring charts leads to prize-giving for the best animal or the owner who has made the most improvement from the start, which is very motivational for the group.

There are many other ways in which the Animal Welfare transect walk (T22) monitoring charts could be compiled, and you should enable the group to analyse their progress in any ways that they choose. Examples include:

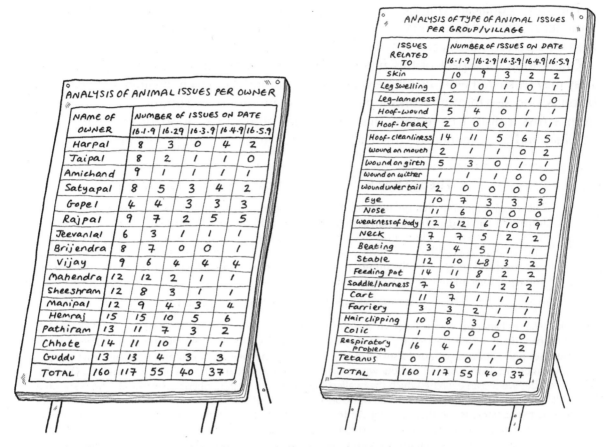

Figure 4.13a (left) Analysis of animal welfare issues affecting each animal owner
Figure 4.13b (right) Analysis of number of animal welfare issues affecting the whole group or village

- changes in animal welfare according to season;
- changes in welfare issues affecting different parts of the animal's body;
- changes in physical welfare issues compared to mental welfare issues.

Case Study I on page 118 gives more information about this analysis.

The impact of the animal welfare improvement process on their own lives

The third area for evaluation is how the group's action to improve their animals' welfare affected their own lives. This is very important because it provides continued motivation to improve animal welfare. You can use tools such as the Before-and-after analysis (T11), Mapping (T1) and Seasonal analysis (T6) to look at specific impacts and changed situations.

Examples from our programmes include:

- reduction in the costs of animal treatment and the effect of this cost saving on the household budget;
- the amount of unity amongst members of the group;
- increased collective bargaining power with resource- and service-providers or decision-makers;
- availability of credit at low interest rates;
- increased self confidence;
- increased recognition in society;
- ability to meet emergency financial needs and the costs of children's education.

The 'most significant change' technique (see Theory box 9 below) can help people to draw together stories and field evidence of their success. This evaluation tool brings out how group members perceive their achievements and the impact on their lives.

One important aspect of this phase is that the lessons learned are shared more widely. They may be shared with the whole village or community to which the animal owners' group belongs, and also through workshops where different groups or communities come together from across a district or region. This encourages others to support or take part in similar activities and helps to increase the reach and effectiveness of your programme.

Theory box 9. 'Most significant change' technique

The 'most significant change' technique was developed in 1996 by Rick Davies in Bangladesh. It is a form of monitoring and evaluation carried out by the animal owners themselves. The technique does not use predetermined indicators of change. Instead the animal owners collect stories of change in their animals and in their own lives a a result of the action they have taken. The stories are read, discussed and analysed by the group, increasing their realization of their own achievements and the impact of their animal welfare interventions.

A 'most significant change' story from India

'Our village, Faridpur, was selected in the summer of 2008 by the Aligarh District Equine Welfare Unit of Brooke India', recalls Hari Singh, a donkey owner from the village. 'Their facilitator conducted several exercises and activities with us to sensitize us concerning the welfare of our animals. Later that year, in November, the team showed us how we could measure the welfare of our animals ourselves. On 27th December we did an exercise which they called 'participatory welfare needs assessment'. We went on a walk through our village and scored all our donkeys using Traffic Light colours. This exercise was an eye-opener for us because we thought that we took good care of our animals, whereas the exercise results showed the gaps in our practices. This led to intense discussion among us because some people did not agree with the results. The Aligarh team told us that the purpose of this exercise was not to highlight our shortcomings but to encourage us to take immediate corrective action for our animals.'

'On 24th February 2009, we did the second participatory welfare needs assessment walk and only one animal out of 22 got 100% score. The third walk was conducted in July and this time seven of the 22 achieved 100%. This time we decided to give prizes for the most improved animal, to encourage and motivate us all to continue improving the well-being of our animals'.

Other donkey owners expressed their gratitude to the Aligarh team by saying: 'This one exercise has induced a belief in us that if we work hard, we can definitely improve the well-being of our animals'.

Source: Monitoring and Evaluation team, Brooke India, 2009

For detailed information on the MSC technique, see Davies, R. and Dart, J. (2005) *The Most Significant Change Technique: a guide to its use.*

Available online at www.mande.co.uk/docs/MSCGuide.htm

Case study I. Using repeated Participatory welfare needs assessment and Root cause analysis to address difficult welfare issues

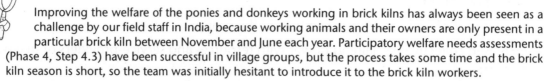

Source: Girjesh Pandey, New Public School Samiti , Amar Brick Kiln, Unnao district, India June 2009

Improving the welfare of the ponies and donkeys working in brick kilns has always been seen as a challenge by our field staff in India, because working animals and their owners are only present in a particular brick kiln between November and June each year. Participatory welfare needs assessments (Phase 4, Step 4.3) have been successful in village groups, but the process takes some time and the brick kiln season is short, so the team was initially hesitant to introduce it to the brick kiln workers.

Despite this, we initiated the PWNA process in Amar brick kiln in January 2009. Mohammed Khalik, our community organizer, visited several times to mobilize the sixteen animal owners working and living there. They formed a group for that particular brick kiln season and set up a savings scheme as an initial activity to bring people together. With this they decided to buy animal feed in bulk and distribute it amongst the members.

Mohammed used the 'If I were a horse' tool (T17) to help the group to identify 28 parameters for assessing animal welfare. They carried out their first Animal welfare transect walk (T22) in January 2009. All the animals were assessed jointly by the group and they made a traffic-light monitoring chart. The parameters for all the animals together helped the owners to get an indication of major welfare issues in Amar brick kiln. The group then did an Animal welfare cause and effect diagram (T26) for the most common and severe issues, such as eye problems, wounds on the withers and spine, cleanliness of the feed trough and maintenance of carts. With Mohammed Khalik's support, the group prepared a comprehensive action plan. Each owner implemented their own plan of action, helping each other and keeping a close eye on everyone to ensure that they were all doing what they had agreed.

The next meeting was held in March: the traffic-light chart was brought out, another animal welfare transect walk was done and the results reviewed. The group concluded that among all the animals, 76 welfare issues out of the initial 154 still existed. They continued to implement their action plan and assessed the animals again three months later, in June. A lot more issues had resolved, but there were still 42 welfare problems left in the kiln.

ANALYSIS OF ANIMAL ISSUES PER OWNER

OWNERWISE: NUMBER OF ISSUES ON DIFFERENT DATES

OWNER	12·1·9	26·3·9	29·6·9
Sanjay	0	0	0
Bisambhar	14	3	3
Mitrapal	14	6	2
Amar Singh	10	3	1
Rakesh	11	6	1
Jaipal	10	6	1
Harpal	10	6	1
Manpal	9	7	4
Pinku	12	6	3
Prempal	12	5	3
Kishan Pal	10	6	3
Ram murat	12	4	9
Ghaiyalal	13	6	5
Arphan	17	12	6
Ganga Ram	0	0	0
TOTAL	154	76	42

Figure 4.14a Analysis of animal welfare issues affecting each owner in Amar Brick Kiln

Two summaries had been made on charts, one showing the number of animal welfare issues per owner (Figure 14a)), and the other containing the total number for the whole group (Figure 14b). Members decided to have a special meeting to look at the two charts and decide on action to overcome their remaining issues.

During this meeting the group carried out a more in-depth Root cause analysis for the unresolved cases, particularly wounds. This brought up some causes for wounds which had not come out during the initial analysis, so a new action plan was made to tackle these causal factors. Each owner agreed to continue to implement the action plan when the brick kiln season ended in June. Several owners were so enthusiastic that they organized themselves into a group at their home village and carried on with the same process of action and reflection.

Facilitating animal owners through this learning process is essential in order to address chronic or long-standing welfare issues which are not easily resolved by the early phases of collective action. The combination of peer pressure for action and good facilitation to analyse and reflect on difficult problems in more depth can lead to improvement in welfare problems which may initially seem challenging or impossible to change.

ANALYSIS OF TYPE OF ISSUES PER GROUP/VILLAGE

ISSUE WISE: NUMBER OF ISSUES ON DIFFERENT DATES

ISSUES RELATED TO:	12·1·9	26·3·9	29·6·9
EYE	24	2	3
EAR	0	0	0
NECK	4	2	1
NOSE	7	2	0
CHEST	8	3	0
STOMACH	4	3	0
HAIR	6	2	1
LEGS	10	3	1
WOUND ON WITHER AND SPINE	11	3	3
BACK	9	3	1
HOOF	2	0	2
CART	36	32	24
FEED MANGER	29	20	6
TOTAL	154	76	42

Figure 4.14b Analysis of animal welfare issues affecting the whole of Amar Brick Kiln

Step 6.2: Gradual withdrawal of regular support

The purpose of this step is for the group to agree on a transition from your facilitation at meetings to a situation where they continue to meet and take action to improve animal welfare without your regular support. This will involve careful planning. As well as individual and collective action by group members, the potential for collaborating with other groups or agencies may come into discussion here. For example, the group may consider furthering their links with local resource- or service-providing agencies.

In the context of an externally-facilitated, intensive intervention to improve animal welfare, withdrawal could mean that you and your organization will no longer support the group with a budget or significant time commitment. After your withdrawal, some contact by you or your supporting organization may need to continue into the future. This might include your attendance at the group's annual gathering or community event, or your support with linking several local community groups into a federation (see Chapter 5).

The decision to withdraw should be based, as far as possible, on the group members' assessment of their own self-reliance and desire to continue to act on improving working animal welfare. This discussion can be facilitated in two ways:

1. Either you can initiate discussion about your regular support during the group's revision of their community action plan, based on the self evaluation described in Phase 6 Step 6.1. Explore the possibility for you to reduce your visits gradually, so that the group can carry out all of their work without you. In their revised plan, participants should identify all the welfare issues for which they need your presence. Make an agreement about how they will deal with some of the issues themselves and also agree on specific deadlines for your input. Incorporating indicators for your withdrawal into the revised plan will enable the group to keep on the right track.

2. Or the group can carry out an assessment of its self-reliance. This can be done through a discussion of what would need to be present in the village for them to be able to continue on their own. The items listed can be used as a checklist for their progress towards self-reliance. Case study J shows an example of this process.

By the end of the withdrawal process, you would aim to respond only to requests for support in a crisis situation which the group cannot resolve on its own.

Although it is possible for a group to carry out this self-assessment on their own, it is even more effective if done through a cluster of welfare groups as seen in the case study. Creation of clusters of animal owners' groups at levels beyond their immediate neighborhood increases the likelihood that each group will continue to work to improve the welfare of their working animals after your withdrawal. The formation of clusters is described in more detail in Chapter 5.

Process box 7. Animal friends

In several of the communities where we work, individuals are selected by the group to be 'animal friends'. They are often active members of the welfare group, with a specific interest in animal welfare. Animal friends form a link between the group and service providers such as local government veterinary practitioners, farriers and feed sellers, and may also be trained by the support agency in basic animal first aid. They stimulate enthusiasm and action by the group and lead processes such as participatory welfare needs assessments. As time goes on, animal friends can take over some of the roles of the facilitator, which makes the process of withdrawal easier.

Case study J. Agreeing indicators for withdrawal of an external agency from regular involvement with an animal welfare group

Source: Kamalesh Guha and Dev Kandpal, Brooke India, Saharanpur, Uttar Pradesh, India, June 2009

In Fakirabad village in Saharanpur district, Brooke India has been working with a group of horse and donkey owners for the last three years. They formed an equine welfare group, prepared village intervention plans and implemented these successfully. Over the years the group has taken up several welfare issues collectively and members have also supported other animal owners to form their own groups. Fakirabad's successes include:

- One of the group members has become an Equine Friend. He is the contact person between the group and local service providers, veterinary health providers and other outside agencies. The service providers are also linked to each other.
- The number of veterinary emergency cases has reduced.
- Owners have become capable of preventing the welfare problems which used to trouble them, such as preventing tetanus through a collective vaccination programme.
- They have established a collective fund which has grown to 50,000 Indian rupees. This is used to support the needs of their animals, such as treatment costs, vaccination and timely repair of carts and harness. Loans from the fund are also used for domestic family needs.
- Group members conduct their own meetings independently, without the presence of Brooke staff or other supporting agencies.

During a recent monthly meeting, the group invited the Brooke's district staff and discussed full withdrawal of their support. This would allow Brooke India to start working with other villages in the area instead of coming to Fakirabad so regularly. Group members asked the question: 'How will we know whether we, or any other group, are ready for the withdrawal of a supporting agency?'

Fakirabad's group leader took this question to a district level meeting of representatives from several different village animal welfare groups. They discussed how they would ensure continuity of their welfare-promoting activities, so that when the Brooke's staff were no longer present, the groups would still be able to run all their activities on their own. With the support of a facilitator, group representatives developed a list of monitoring indicators for withdrawal (see Theory box 10). This indicated what needed to be present in any village or group to ensure it was ready for withdrawal of a supporting agency. Then the group of representatives decided to visit each village along with field staff from the Brooke. They used the list of indicators to look at animals, meet with group members and assess their meeting records and registers. They also met local veterinary workers and other service providers. During each visit, scores were given against each indicator and these were shared with the local group members to build consensus. When all of the villages had been visited, a meeting was held to decide which ones were ready for gradual withdrawal of support. Five villages were chosen, including Fakirabad. It was agreed that the process of withdrawal would be gradual, so a Brooke community facilitator still visits Fakirabad occasionally and the group can still ask for help whenever there is a clear need.

Theory box 10. Indicators for withdrawal of an external agency from regular involvement with an animal welfare group

1. The group is functioning well with regular meetings, and any conflicts or problems are resolved collectively and recorded in village registers.
2. Community-led tetanus vaccination is being carried out regularly, including booster doses and vaccination of newly-purchased animals.
3. The Animal Friend plays an active role and all members are confident in his work.
4. There is a collective understanding of the major local diseases affecting horses and donkeys, their symptoms and prevention, including wounds, laminitis, colic, eye disease and tetanus.
5. Local veterinary service providers are available for treatment of animals, and their services are being used by the community, including taking regular advice and getting sick animals treated on time.
6. A first aid system has been arranged involving the local medicine retailer and this is being used by the majority of the group.
7. A farrier, hair clipper and feed seller are available locally and provide good services to the satisfaction of all group members
8. Men, women and children in families who own animals are all aware of the animals' welfare issues and participate in activities that are important to improve and maintain good welfare.
9. Collective action is being taken by the group to meet their animals' needs, such as feeding and prevention of diseases, and these are recorded in the village register.
10. Shade and shelter is arranged for working animals according to the weather.
11. Arrangements are in place to provide good quality water to animals frequently.
12. Stables and manger are properly cleaned, and animals are provided with appropriate feeding, such as wet wheat straw.
13. Group members have a good knowledge of cart balance and saddle-fitting and there are no sharp edges on the bit.
14. Animals and owners show calm and friendly behaviour towards each other.
15. Wounds are being managed properly and animals have few or no wounds.
16. There is no lameness in the animals.
17. Animals are being groomed and their eyes are being cleaned regularly.
18. There are no signs of firing (branding with hot irons) on the animals.
19. Male animals' genital organs are not being tied with string to prevent normal behaviour.
20. Animals do not start to work at an early age which damages their development and increases lameness.

CHAPTER 5
Reaching out to promote animal welfare

What you will find in this chapter

This chapter explores several ways in which collective action to improve animal welfare may be extended to reach larger groups of people and populations of working animals.

Building on your work with action-oriented community groups in Chapter 4, the first part outlines how to bring these groups together in order to maintain a long-lasting improvement in the welfare of their animals.

The second part looks at methods for raising awareness of animal welfare and engaging with audiences who are not based around established community groups. In our experience these methods are less effective for promoting sustainable, collective action than intensive engagement with community groups. However, they are very useful for stimulating interest in animal welfare among wider society and appreciation of the important role of working animals in people's lives.

There is a large amount of information available on the use of community outreach methods in different fields of international development. In this chapter we outline our own adaptation of these methods for promoting good animal welfare with communities and larger groups. When you have identified the best methods to use with your target audience, we recommend that you read some of the references relating to those methods in more detail. Then use your field experience and the animal welfare information in this manual to generate your own outreach activities and communication materials.

Forming federations

Step 1 Building opinion and forming a federation 0 to 6 months

Purpose
- To organize animal owners from different local groups to come together, in order to form a higher level solidarity group or federation to address the common animal welfare interests of its member groups.

Process
- Build opinion on the prospects for federation, through meetings and discussion at group level.
- Bring understanding to each group on the role of federations and their own roles as members, through workshops and training.
- Devise an appropriate process for selecting group representatives to the federation, followed by orientation for representatives.
- Develop federation goals and objectives with the member group representatives.
- Facilitate: framing of rules and regulations; agreeing clear functions, roles and responsibilities of the federation and members; devising activity plans for the federation.

Step 2 Stabilising the federation 6 to 24 months

Purpose
- Regular facilitation and support to build the capacity of representatives to manage the newly formed institution, run it properly and grow it effectively.

Process
- Enable federation members to put their activity plans into operation, in accordance with their goal and objectives.
- Capacity-building of members in subjects such as managing their action plan, resolving conflicts, organizational development and any specific skills needed to carry out their work.
- Enable federation members to identify like-minded institutions, build links and network for benefit of the member groups.
- Help establish records, registers, financial management systems and audits.

Step 2 Stabilising the federation 6 to 24 months

Purpose
- Slow withdrawal of direct support and facilitation once the federation becomes effective.

Process
- Periodically attend federation meetings.
- Support and facilitate the federation strengthening process until it reaches a sustainable level.
- Participatory impact evaluation and re-planning of activities.
- Attend meetings and events occasionally to see that the federation is continuing in the right direction, while giving active support only in extreme need.

Table 5.1 Forming a federation of working animal welfare groups

Facilitating the establishment of working animal welfare federations or intermediate-level solidarity groups

Creation of collectives of animal owners' groups at levels beyond their immediate neighbourhood is an important method of sustaining momentum after you (the facilitator), your organization, or other external forms of support are withdrawn.

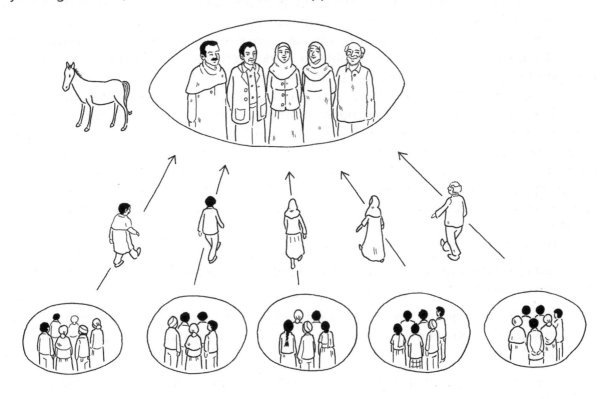

Why federate local community groups?

Reasons for bringing community groups together as animal welfare solidarity groups or federations include:

- Enabling groups to take up actions which would not be possible if they acted alone. An example of such action might be to encourage local authorities to improve services or resources which improve the welfare of working animals, such as the provision of animal water supplies and shade shelters in a town centre.

- Sharing and learning: federated groups have greater opportunities for sharing animal-related information, so that they can sustain their own activities more effectively. This may be achieved through creating an animal welfare platform (regular gathering) of group representatives who bring relevant, useful, and interesting information to these assemblies and take learning back to their own communities. Local or district newsletters may be used for the same purpose.

- Federations can also form a platform for alliance-building and networking with other stakeholders, in order to build a collective appreciation, opinion and good image of working animals among wider society. Stakeholders may include groups of animal users or hirers who do not own animals themselves, non-government organizations, local administrators, higher institutions of learning, the wider community, schools, churches, local leaders, community opinion-leaders, community-based organizations, agro-business practitioners, artisans and other interested parties.

How to form federations

The process for forming a federation of working animal welfare groups is laid out in Table 5.1. At the start, active members of several village groups may be brought together to discuss the opportunities for forming a cluster group.

Managing a large federation may not be possible at the beginning, so we suggest that you start with facilitating five to fifteen animal-owning groups to form a cluster. The formation of a cluster or federation is more effective if the groups who join are well established and active (see Theory box 10). Unless the groups who join have reached a stage of self-management and self-motivation, they will not be able to work effectively in the cluster. Each group can send two or three representatives to an assembly event. At the beginning this should focus on sharing and learning about each others' achievements, giving examples of the specific results and impact of interventions on the welfare of working animals. The event may take one day or longer, depending on the number of the groups who are meeting. Such events can also include competitions, as described in the case study below. Cluster meetings can then be organized on a quarterly basis or at time intervals agreed by the group representatives. Cultural events such as folk songs and community drama may be held during cluster meetings in order to encourage representatives to interact creatively and sustain their enthusiasm.

Factors to consider when organizing cluster meetings or events include:

- distance which group representatives would need to travel in order to meet other cluster members

- similarity of the groups' working or livelihood context (for example, rural animal owners may see their situation very differently to people working in brick factories or in urban areas)

- specific working animal welfare issues which the cluster or federation might decide to address

Theory box 11. Conditions necessary for a community group to become a successful member of a cluster or federation

- Group meetings are held regularly and attended by all members
- A common fund has been built up through regular contributions or savings by all members and it is used for collective and individual actions
- Rules and regulations are well framed and properly governed
- Records and registers are maintained properly
- Leadership roles are rotated

A federation needs to build up its own fund to carry out its agreed action and implement collective decisions, for example to maintain records, register as a cluster with the local authority, hold meetings or travel to meet officials. These federation funds can be built up in several ways:

- Membership fees or admission fees as a one-time payment from member groups

- Monthly contributions from member groups

- Donations or contributions from other institutions for programmes within the scope of the cluster's objectives

- Money earned from any income-generating activities taken up by the cluster, such as income generated though bulk buying of animal feed and sale of feed mixtures to cluster members.

- Fines from member groups, for example for non-attendance or other violations of the federation's rules

The cluster could provide a monitoring role for the functioning of individual groups, through regular review of the work of member groups. It may also try to strengthen community groups through ideas, suggestions, exposure visits, audits and training. Activities which could be carried out by the member groups themselves should not be taken up by federations – it is more useful for them to concentrate on animal welfare issues which require a higher level of collective action.

Case study K. A village-to-village competition

Source: Dev Kandpal and Dinesh Mohite, Brooke India, Saharanpur, Uttar Pradesh, India, October 2008

Brooke India's equine welfare unit in Saharanpur district started community mobilisation activities in about 30 villages where horses, mules and donkeys were used for work. For the first year, the Brooke's staff focused their activities on facilitating animal-owning families to form and strengthen 'equine welfare groups' with regular savings and loan schemes as the activity that initially brought each group together. These groups also began collective action on various welfare issues affecting their animals. Each village made a specific welfare improvement plan, which was monitored jointly by the group members and the Brooke every month. During the first two years, progress was found to be very satisfactory in each village. Planned action by the groups improved the welfare of their animals, service providers for animals were identified and were influenced to provide better quality and more cost effective services. 'Equine Friends' in each village worked as a link between owners and service providers.

At this stage, it was decided to hold a meeting of all these equine welfare groups together, bringing together two or three representatives from each group to share and learn. Everyone decided that a prize would be given to the group who best encouraged all the other groups. All the representatives sat together to decide criteria for judging the competition. These included the amount of money saved and loaned by the group, the number of animal welfare problems sorted out collectively, the number of animals vaccinated and the reduction in veterinary emergencies affecting the group. Other criteria included helping people to form groups in other villages and enabling other groups to sort out critical problems. After six months the representatives met again to look at the results from each village and to distribute prizes. This created a high level of competition between village groups and as a result positive changes in animal welfare started to happen very rapidly. All the animal owners were charged with enthusiasm and jumped to take part in collective action, both within the group and outside it.

Now we see that collective meetings are not confined to the Saharanpur district but have crossed into other districts. Groups from three or four districts organize their own meetings, deciding on more collective action to help their animals. Participants write and sing their own folk songs about horse and donkey welfare, act out community drama and distribute prizes. Competitions continue independently between the districts and the Brooke's staff are invited to participate. These competitions empower people and help the groups to celebrate and learn from each others success. They also mobilize and motivate groups towards simple and concrete actions and focus them on fulfilling a future vision for their animals.

Methods for extending reach: community outreach

It may not always be possible to work with animal-owning community groups using the methods described in Chapter 4. In rural areas, working animals and their owners may be too scattered geographically to form effective groups or platforms for community action. In peri-urban or urban areas, the population may be too large to work with everyone at the same time. Sometimes practical constraints make it difficult to meet animal owners in their home villages. We have found it quite challenging to work with animal owners and users at congregation points during their working day, such as at markets and caleche or tonga (horse-drawn taxi) stands, because they have little time and opportunity there for group formation or reflection. In places like these, awareness-raising approaches can provide a cost-effective method of outreach to improve working animal welfare. They are usually less effective than formation of community groups, because increasing individuals' knowledge and awareness does not necessarily lead to a change in their behaviour towards their animals. However, these more extensive approaches can still be useful, particularly when animals are working in environments which pose fewer risks to their welfare (see Chapter 3, 'Deciding how to work: the intervention approach').

An example of community outreach is the wall-writing carried out by the owners of working animals in Uttar Pradesh, India. These owners are often scattered in many villages and there is limited opportunity for them to meet each other and form cohesive self-help groups. To raise awareness of animal welfare needs, the owners decided to write their good animal husbandry and management practices on the outside of their stable walls. This educated other animal owners as well as showing pride in the fact that they were caring for their animals well.

In some countries, such as Kenya, many welfare problems are associated with the low status of working donkeys in society. Raising awareness of their value and role in supporting livelihoods, by spreading animal welfare messages to society at large, may be an essential part of your approach in such circumstances.

Some of the extensive methods described in this chapter may also be used during the early stages of the group formation process and feeling the pulse of the community (see Chapter 4) in order to raise awareness about specific welfare issues to be taken forward for collective action.

Process box 8. Key principles in community outreach

- Develop a plan for outreach and avoid ad-hoc messaging, which will not be successful.
- Involve people as actively as possible, by choosing the method of outreach with the highest level of participation by the target group.
- Use a variety of channels and methods to provide space for people to express themselves and their concerns about their working animals.
- Communication should not be based on assumptions about what people know and don't know, so carry out a proper analysis of information needs (for example using Practice gap analysis, T21).
- Develop outreach materials locally wherever possible.
- Involve animal owners in creation of the media and messages.
- Ensure there are opportunities for feedback and monitoring by the animal-owning community.

Your community outreach plan should be clearly defined:

- What would you like to communicate?

- Who would you like your message to reach?

- How would you like to communicate it?

Then the community outreach plan needs to be integrated into your wider project or programme.

Although your contact time with the target group of animal owners may be limited, or you might meet them irregularly, it is still important to use methods which generate two way communication, discussion and dialogue among the participants. Your challenge as the outreach facilitator is to stimulate interaction with your audience, sharing of experiences and learning between them, and to involve animal owners, users and their families as much as possible. One-way communication using posters, booklets or leaflets alone is not enough to stimulate behaviour change to improve animal welfare. The success of your outreach depends on the involvement of the people for which the communication is designed. All media give scope for participatory input and most can be successfully combined with other participatory methods including those described throughout this manual.

Which of the illustrated scenarios below do you feel is more effective, and why?

This...   *... or this?*

Given your relatively limited direct contact with the communities that you are aiming to reach, strategies for monitoring and evaluation are challenging. However this does not make them less important. Several ways are possible, such as:

1. Embedding an element of feedback into the design of extensive communication approaches. This can be easily incorporated into the design of interactive media like community theatres and other outreach methods based on creative performance. Selective assessment of how messages delivered by non-interactive media – such as radio spots or television features – have been understood to also be an effective way

to monitor and evaluate communication packages. For example, audiences could be asked to write or phone in with their feedback and suggestions.

2. A select group of sensitive and interested animal owners identified through the intensive programmatic approach (see Chapter 3) could be involved in the design of extensive strategies. Their expectations and suggestions may be used as monitoring and evaluation indicators for the extensive approach. A working group of such owners could be called upon every now and again to comment on the design and content of outreach communications.

3. Make use of every fieldwork opportunity to understand the reach of extensive approaches and communication materials and the perceptions of the animal owners who receive them. It is important to reach out pro-actively to different groups of animal owners, users and carers in order to hear their views about your communication and whether it has affected their animal management and work practices.

Channels of communication for outreach in communities

There are many different potential channels for community outreach. When you are working in the field, identify the channels or opportunities which are most suitable and specific to the communities with which you work. Discuss this with communities and look around you to see when and where they get different kinds of information that is useful to them. Examples from our experience include the following.

Work congregation points such as market places and *caleche* or *tonga* (horse-drawn taxi) stands

Special occasions

In some countries working animals are used during special occasions, such as religious pilgrimages and festivals in India. These may provide an opportunity to set up animal welfare camps or events together with a local organization or institution. For example, you may work with the government veterinary department to provide outreach on animal health and welfare practices.

Social gatherings such as water points

Exhibitions, fairs and field days

These could be fairs where working animals are traded or annual exhibitions organized by the local authority on trade and agriculture. Exhibitions and fairs can be used as a platform for sensitizing animal owners and users, traders, government and wider society to welfare themes and good practices.

Methods of communication for animal welfare outreach

There are many methods of communication that you could use to convey messages about the welfare of working animals. Together with animal owners and users, choose the best method for your local context. Wherever possible, involve them in creation of the media and messages. We do not have space in this manual to include a large amount of information about different communication methods, so further information is signposted in the further reading and reference list at the back of this manual.

Case studies from our own work include:

Contests and competitions

These can be very effective events in engaging animal owners and users, whether they are children, youth or adults. The key to effectiveness is people's full participation in deciding the criteria for winning and then selecting the best-kept or happiest working animal. Competitions can be organized within a village or between villages (see Case study K).

Traditional cultural activities and drama (Theatre for Development)

This group of effective methods includes talking, storytelling, song, dance, theatre and puppet shows. There are many ways of using these to create awareness about animal welfare, either by involving working animals directly in the drama or by using the drama to enact and stimulate discussion on a local animal welfare issue.

Community drama enables animal owners, users, carers and other stakeholders to participate by outlining their fears, needs and aspirations about their working animals. Here are some examples.

- In Pakistan a group of garbage collectors formed an acting group to write and develop sketches about donkey welfare for other owners.

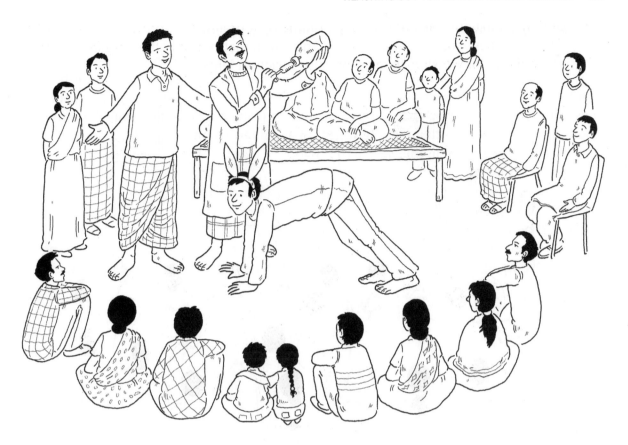

- In some of our groups, individual owners write songs and poetry about working animals and share them as part of their group meetings.

- In India and Egypt, local puppeteers have developed puppet shows about donkey welfare for children and performed them at animal fairs.

- During exhibitions, fairs or field days, drama by a professional group may be used to raise awareness about animal welfare. This can target not only the people who own, use and care for working animals, but also society at large.

Many good books and documents about community drama are readily available; some of these are mentioned in the further reading and reference list at the back of this manual.

Recorded songs or stories are often very popular and can be played using low cost, low-tech communication systems such as car batteries or a generator. During meetings, public gatherings, fairs and exhibitions, recorded materials can be used as part of interactive discussions in large forum. These recorded programmes can focus on specific animal management practices or cover more general welfare themes. Material can be recorded with the help of professional teams, or by the community itself. An example of community-generated recording is the use of **Participatory Videography**, (see the further reading and reference list for a good manual on participatory video).

Radio provides another way to disseminate messages about working animals to their owners, users and wider society. In many countries, transistors or FM radios are carried by the people who borrow or hire working animals to earn a living. Radio is a potential method for reaching this group of people, who may otherwise be difficult to engage in discussions about animal welfare. See Case study L for an example of using radio to promote animal welfare. Community radio is a relatively new development in which the community itself develops the radio programme. More information on radio is included in the further reading and reference list.

Case study L. Radio messaging to improve the welfare of working donkeys

Source: Heshimu Punda (Respect Donkeys) Project, Kenya Network for Dissemination of Agricultural Technologies, 2006

The Kenya Network for Dissemination of Agricultural Technologies or KENDAT – a Kenyan non-government organization started a donkey welfare radio programme in April 2003. The *Mtunze Punda Akutunze* radio project began airing in Kiswahili on a local radio station. *Mtunze Punda Akutunze* is the Kiswahili translation for 'Look after your donkey and she will look after you'.

The main objective of the project is to design and produce innovative radio messages to improve the use, welfare and work environment of donkeys carrying out transport and tillage in rural communities. The programme has educational episodes and at the end of each episode there is a question to gauge people's understanding of the topic of the day. Listeners are urged to send in their answers and winners are rewarded with t-shirts each week. They are also encouraged to send in questions about their donkeys and the radio programme teams reply on air to all letters received. Today the donkey welfare radio programme has reached national radio and is broadcast on the Kiswahili radio station Citizen Radio at the prime time of 8.30 pm on Saturdays.

The wide range of donkey welfare issues covered by Mtunze Punda Akitunze includes:
- Awareness-raising and attitudinal change on the importance of donkeys and their welfare.
- Donkey diseases and their prevention.
- Handling and caring for expectant female donkeys and their foals
- Proper harnessing to reduce wounds and sores.
- Donkey health including hoof trimming, de-worming, wound dressing, treatments.
- Donkey use and management such as tillage and transport, housing, tethering, rasping of teeth.
- Nutrition and feeding including types of feed, quantities, watering and supplements.

The educational episodes cover different topics and are presented in an interactive way through interviews or discussions, engaging various stakeholders and experts on issues pertaining to donkey welfare. Feedback from listeners is an important part of the programme and an indicator of the level of interest of the listeners. The radio programme has recently hired a scout who goes around the country and gives awards to the owners of donkeys that are working in an outstanding welfare environment. It has also encouraged formation of fan clubs composed of people who are enthusiastic about donkey welfare, who come together as a community or interest group to discuss welfare issues. Such clubs are announced over the radio and awards are presented to encourage membership. This project has demonstrated the power of the portable radio in Kenya.

Posters, leaflets and newsletters

Posters should be used for brief animal welfare messages and work best if the message is visual. Leaflets may be used to communicate more technical information. Make sure that you know the literacy level of your target group when making leaflets. Many owners of working animals are not literate so written leaflets may not be best for them, although they could be very effective in promoting good welfare to agricultural extension staff (see example of an information poster for extension staff in Figure 5.1 on page 138).

Newsletters can be useful for literate audiences. We have used English-language newsletters in Ethiopia to promote awareness of animal welfare amongst partner organisations such as government agencies and non-governmental organizations. We have also produced newsletters in India, written in Hindi, which are used for a variety of reasons including:

- sharing success and learning between animal-owning groups within a district;
- inspiring and developing a competitive attitude between these groups by providing evidence and stories of success;
- providing public recognition of success which increases the self-esteem of group members;
- providing technical information and promote indigenous technical knowledge.

Tetanus
(Hawa ki bimara/Hironbyal/Jakda- local language)

Cause: Bacteria present in soil and animal dung, the bacteria enter into open puncture wounds and spreads out toxins which cause the disease

Symptom:

- Stiffness and rigidity of the head and neck
- Raised tail and erect ears
- Flared nostrils and surprised look
- Third eyelid easily visible
- Horse moves stiffly as if it is made of wood
- Extreme sensitivity to sounds, light, touch

Prevention:

- Prevent puncture wounds by nails, wire and other sharp things
- Clean wounds carefully and leave them open to air
- Vaccinate (2 injections 4 weeks apart) then single booster every year
- Breeding mares should be vaccinated 4-6 weeks before foaling

Note: Prevention is better than cure – VACCINATE! Treatment is very costly and often animal does not recover and can die!

 Brooke Hospital for Animals

BROOKE

Figure 5.1 Example of an information poster for extension staff

Theory box 12. Checklist for developing effective communication materials to improve the welfare of working animals

Try to involve the animal owners, users and carers in all steps of the process, even in the definition of the purpose and objective of the proposed communication. This will increase its effectiveness.

1. Based on your purpose and objective, define the aims of the communication.
2. Identify, define and prioritise the audience and participants.
3. Determine what information they would like to know.
4. Identify which sources of information the people normally use.
5. Identify barriers and opportunities for good communication.
6. Identify the most suitable communication channels.
7. Develop messages – participatory involvement in message formulation is important, in order to create specific messages for specific target groups.
8. Plan coordinated timing of activities.
9. Formulate communication materials.
10. Pre-test the materials in a participatory way.
11. Implement use of the communication materials.
12. Participatory evaluation.

Adapted from Burke (1999), 'Sample Communication Strategy' in: *Communication and Development, A Practical Guide*, p. 25.

PART III
PARTICIPATORY ACTION TOOLS FOR ANIMAL WELFARE

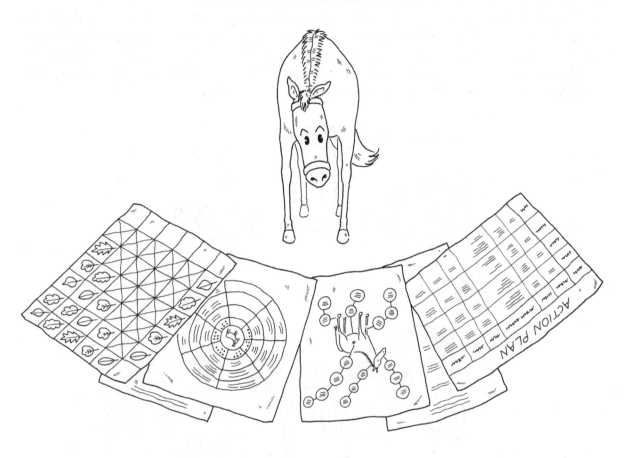

What you will find in this toolkit

This toolkit contains participatory action tools for animal welfare. Some are adapted from existing participatory rural appraisal (PRA) tools. Where no appropriate tool existed, or where adapted versions did not work well when tested with communities, we developed new tools specifically designed for the purpose of facilitating communities to understand, analyse and improve the welfare of their working animals.

All the tools have been extensively tried and tested in the field by our field facilitators, who have created many variations to meet the specific needs of animals and their communities. In this toolkit the versions that are most commonly used are presented and at least one alternative or variation per tool is provided. However, we recognize that no tool will be applied in exactly the same way each time and there is potential for immense creativity and innovation by field facilitators. We hope that you will continue to adapt and innovate in order to find the most effective ways to enable communities to improve animal welfare.

Expectations when using this toolkit

This is not a comprehensive manual to train you to use PRA tools. It requires you to have a basic knowledge of PRA tools for analysis of space, time and relationships, as described in many other books such as *Methods for Community Participation* (Kumar, 2002). If you have prior experience in using some of these tools in their original context (to facilitate behaviour change in other sectors such as health education, or water and sanitation), this will give you a deeper understanding of how to use them in the animal welfare context.

How to use this toolkit

For each tool we describe the following:

1. **What it is:** a brief description of the tool and its original use (if adapted).

2. **Purpose:** the areas of discussion and analysis that the tool will facilitate, and its adaptation for use in animal welfare.

3. **How you do it:** the steps you will take to use the tool in an animal welfare context. It focuses on the part of the process that is different from the original tool.

4. **Facilitator's notes:** these are practical hints and tips about each tool, offered by our field facilitators as a result of their experience in using the tool with animal-owning communities.

5. **Figures and examples:** real examples of the tools in use with animal-owning communities have been copied from field documents and translated to aid your understanding.

If you are an experienced facilitator with your own community mobilisation process, you may wish to find a tool for collecting specific animal-related information for discussion. A symbol (icon) next to the description of each tool shows the type of discussion for which the tool is used:

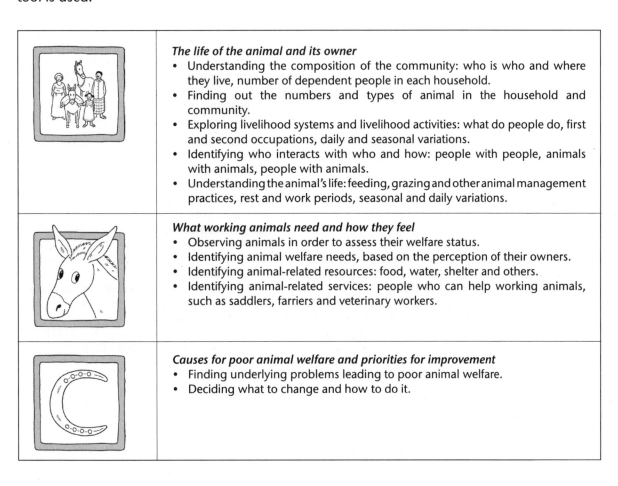

	The life of the animal and its owner • Understanding the composition of the community: who is who and where they live, number of dependent people in each household. • Finding out the numbers and types of animal in the household and community. • Exploring livelihood systems and livelihood activities: what do people do, first and second occupations, daily and seasonal variations. • Identifying who interacts with who and how: people with people, animals with animals, people with animals. • Understanding the animal's life: feeding, grazing and other animal management practices, rest and work periods, seasonal and daily variations.
	What working animals need and how they feel • Observing animals in order to assess their welfare status. • Identifying animal welfare needs, based on the perception of their owners. • Identifying animal-related resources: food, water, shelter and others. • Identifying animal-related services: people who can help working animals, such as saddlers, farriers and veterinary workers.
	Causes for poor animal welfare and priorities for improvement • Finding underlying problems leading to poor animal welfare. • Deciding what to change and how to do it.

In the table below, the Participatory action tools for animal welfare are listed according to the community mobilisation process described in Chapter 4 of this book. This may help guide you through the steps if you are unfamiliar with such a process. It is not necessary, and indeed not advisable, to use all of the suggested PRA tools for each step. Using too many tools or exercises at the beginning may create confusion and bad feeling, which results in gradual loss of interest, low attendance and low participation in meetings. Remember, your goal as a facilitator is to start up and maintain the process of creating a climate for collective action. It is not merely the completion of particular tools in a particular way, or in a specified order.

Table T1 Overview of the use of participatory action tools for animal welfare in the community mobilization process

Phase	Step	Tool
1 Feeling the pulse	1.1 Building a rapport with the animal-owning community	Mapping (T1) Daily activity schedule (T4) Gender activity analysis (T5) Animal welfare snakes and ladders game (T16) Historical timeline (T7)
	1.2 Forming and strengthening an animal owners' group	Seasonal analysis (T6) Dependency analysis (T12) Credit analysis (T13)
2 Shared vision and collective perspective	2.1 Identifying issues relating to: (i) the livelihoods and working systems of animal owners	Mapping (T1) Mobility map (T2) Venn diagram (T3) Daily activity schedule (T4) Gender activity analysis (T5) Seasonal analysis (T6) Gender access and control profile (T10) Changing trend analysis (T11)
	(ii) the lives of working animals	Animal welfare and disease mapping (T1) Animal disease venn diagram (T3) Daily activity schedule of the animal (T4) Dependency analysis (T12) Animal body mapping (T20) Animal welfare practice gap analysis (T21)
	(iii) animal-related service-providers and resources	Animal-related service and resource mapping (T1) Mobility mapping (T2) Pair-wise ranking (T8) Matrix scoring of animal-related service providers (T9) Cost-benefit analysis (T15)
3 Participatory animal welfare needs assessment	3.1 Analysing how animals feel and what they need for their wellbeing	Matrix ranking of animal welfare issues (T9) 'If I were a horse' (T17) How to increase the value of my animal (T18) Animal feelings analysis (T19) Animal body mapping (T20) Animal welfare practice gap analysis (T21) Animal welfare transect walk (T22)
	3.2 Generating a list of animal-based and resource-based indicators of welfare and agreeing on how they will be scored	
	3.3 Observing animals and recording their welfare status	
	3.4 Analysing the present welfare status of animals belonging to the group	
4 Community action planning	4.1 Prioritizing welfare issues of importance to working animals and their owners	Historical timeline (T7) Pair-wise ranking (T8) Matrix scoring (T9) Three pile sorting (T23)
	4.2 Root cause analysis of welfare issues	Animal welfare story with a gap (T24) Problem horse (T25) Animal welfare cause and effect analysis (T26)
	4.3 Preparing a collective plan of action to improve the issues	

Phase	Step	Tool
5 Action and reflection	5.1 Implementation and monitoring of activities in the community action plan	Pair-wise ranking (T8) Matrix scoring or matrix ranking (T9) Animal welfare transect walk (T22) Problem horse (T25)
	5.2 Participatory monitoring of animal welfare changes, creating a cycle of reflection and action	Animal welfare cause and effect diagram (T26)
6 Self-evaluation & gradual withdrawal of regular support	6.1 Self-evaluation 6.2 Gradual withdrawal of regular support	Changing trend analysis (T11) Group inter-loaning analysis (T14) Success and failure stories (Chapter 4, Phase 6)

Tools for specific animal management purposes

At the end of the toolkit you will find two examples of how the tools listed above can be used to look at specific management issues. Analysis of animal feeding practices (T27) is a tool which involves both the community and an external expert, such as a veterinarian or animal nutritionist. Village animal health planning (T28) combines several of the tools listed above. We have used these successfully with many communities to improve the nutrition of working animals and to develop plans for disease prevention and control. Meanwhile, more complex variations and adaptations of these tools continue to be developed.

Process box 9. Hot tips on using PRA tools

- Make sure you explain the purpose of the exercise clearly to the group of participants at the beginning.
- Agree on how the outcome of the exercise (chart or diagram) will be shared afterwards: who will see it and how it will be used?

- Get the right group together. Do you want to work with animal owners and their families together? Is it useful to talk with animal carers and animal owners separately? Will more people participate if men and women are in separate groups? Is there a potential social problem in doing this? Try to work with a group of maximum 10 to 15 people. If the group is larger it is best to split it between two or more facilitators. The groups can do the same exercise or different ones and then come together for a discussion.

- Participants may represent their activities using the original items that they use in daily life, such as a broom for the times that they clean, or a food bowl for the times that they eat. Alternatively they may use seeds, leaves, sawdust, lime powder, coloured cards, chalk or any other materials available locally. The exercise can be represented on chart paper or newsprint, although depicting the chart on the ground enables people to see and make changes conveniently.

- Try to show activities using pictures or symbols. This avoids excluding people who cannot read and write. Scoring or weighting can be done with beans and seeds, or sticks of different lengths

- While doing an exercise like mapping, ask uninvolved observers whether or not they think the placement of a particular feature is accurate. If they disagree with the placement, invite them to indicate its proper position. It is important that representatives of different groups within the community are involved in the process wherever possible. Encourage everyone to express their own views.

- If a particular participant is dominating the group, ask him or her specific questions about the village or agree to interview him or her separately later as part of the process. Engaging this person in conversation away from the map or chart will reduce their influence over the process.

- With all PRA exercises, the discussion, reflection and analysis are more important than the maps, charts or diagrams themselves. This analysis can bring the rich and diverse experiences of the group together to build understanding about things of importance to the community.

- Watch and wait for the moment to discuss your points. Do not interrupt the participants' conversation flow. Take notes of all the interactions and things that are discussed.

- Documentation is important so that the exercises can be referred to in future and used for monitoring progress over time. Explore with the group how they would like to keep records of their discussion and activities. Maps and charts can be copied onto chart paper, A4 paper or into a ledger, with a copy for the group to keep and a copy for you as the facilitator. School children or college students from the community are very effective at doing this.

- Include the village name, the date and the names of all participants. Make sure there is a key to show the meaning of all the symbols used.

- At the end, always thank people for their time and their active participation.

T1 Mapping

Animal-related services and resources, animal welfare and animal diseases

When you first start to work with a community, it may be difficult to know where and how to begin. Most community members are not accustomed to being asked for their expertise. Mapping is a good tool to begin with, because it is an easy exercise that gets communication and discussion going between you and the community. Local people are rich in knowledge and understanding about their own surroundings, where their families have lived often for generations, so maps drawn by the local community are usually detailed, authentic and accurate.

What it is

A map is a visual representation of the important places, services and resources in the area, as seen and understood by the community at the present moment. Maps can also show the welfare status of animals or show animals affected by particular diseases or problems. Using this tool will begin to focus both you and the community on the animal welfare problems and issues which need further investigation.

There are many ways in which you can use mapping. Here we have described some variations.

Animal services and resources map

This social map (Figure T1a) shows the community's own village and environment. It includes houses, roads, drinking water facilities, working places and natural resources, as well as resources related to working animals such as grazing land, resting, feeding, and watering areas. It identifies which places and people are important to the community regarding the care of their animals, such as the location of feed sellers, farriers, hair-clippers, cart-makers, local health service providers, agrovets or pharmacies and government veterinary clinics.

Animal welfare map

An animal welfare map (Figure T1b) shows the welfare status of animals in each household, based on specific factors which the community find important. It gives an overall 'bird's-eye view' or an overview of animal conditions in the village. Depending on how the community sees the welfare of its working animals at this early stage, the map may contain a mixture of:

a. Animal-based observations, such as weak and healthy animals, lame animals, animals with wounds or injuries.

b. Behaviour of animal owners, such as who overloads or beats their animal, or doesn't keep the animal shelter clean.

c. Resources, such as who has sufficient space or access to grazing land for keeping animals and who does not.

This map can be used to analyse the current animal welfare situation in the village and to identify the first problems that come to mind (Tools T21 'If I were a horse' and T25 Problem horse do this in more detail). Displaying and discussing the welfare status of each person's animal is the first step in creating peer pressure for change. The map may also be used to identify current animal management and work practices, acting as a baseline for monitoring future progress. When repeated after a period of time, changes in animal welfare status and management practices can be shown on the map.

Animal disease map

This (Figure T1c) shows the animals affected by disease, blindness, lameness and other conditions; it may be used for both present and past disease problems in the village. Showing diseases on a map encourages analysis and further discussion of their symptoms, causes and routes or patterns of infection, as well as their effects on the animal, owner and community. The discussion can explore people's concerns about the health of their working animals and the things that they would like to change.

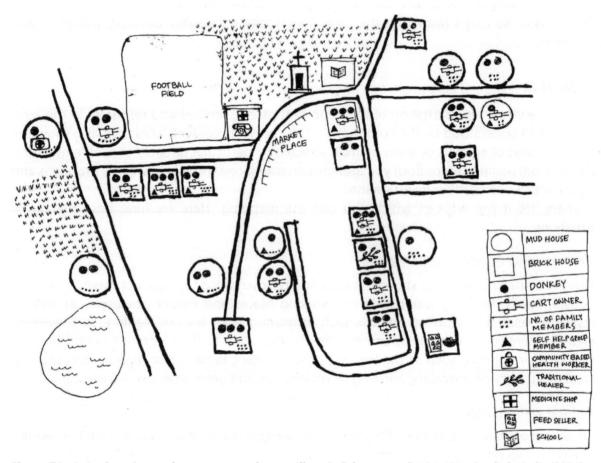

Figure T1a Animal services and resources map from a village in Saharanpur district, Uttar Pradesh, India (2006)

Once a village outline had been drawn, the group added the number of donkeys per household (black dots) and whether the donkeys worked with carts or packs. Animal-related service providers and resources are shown in the legend and include the community-based animal health worker, medicine shop, feed seller and traditional healer. Discussions during and after production of the map focused on the availability, cost and quality of the resources and service providers available in the village.

How you do it	
Step 1	Start by asking the group to draw a sketch of their village site on the ground using any local materials they choose, showing: • general infrastructure, such as roads and paths; • households in the community, including details of each household's family members; • places which are important to them, such as the doctor, temples, the mosque and the places where they hold meetings
Step 2	To make the map more animal-focused, ask the group to indicate each household which has working animals. Show the types and numbers of animals using symbols or local materials. Add details such as whether the animals are male or female and the kinds of work they do.
Step 3	Identify any services and resources for animals which are available in the village: • natural resources, such as animal water supplies, resting areas and grazing areas; • service providers, such as farriers (blacksmiths/shoemakers), cart-makers, feed sellers, animal health service providers, livestock extension workers and so on. According to the discussions among participants, you might wish to ask questions about: • resources and services which are not available in the village, and where participants go to access them; • availability, quality and cost of services, highlighting different people's views; • reasons why some people have access to resources and services and others do not. Later on we describe specific tools to analyse these questions in more depth (such as T12 Dependency analysis). You may introduce them at this point if the group wishes, or bring them in at a later meeting. At this point the Animal services and resources map is complete. Ask participants if there is anything else that they would like to show on the map which they think is important to any aspect of their life or livelihood. Alternatively you may wish to continue to develop the map further, into an animal welfare map or an animal disease map (following the steps below).
For an animal welfare map use these three steps:	
Step 4	Ask participants to discuss and agree which working animal is the best in the village. Mark it on the map using a symbol next to the house. Then ask if any other animals in the village are in a similarly good condition. Give these households the same symbol. Next, identify which animals are in a moderate condition, marking them on the map using a different symbol. Finally, show the animals which are in the worst condition. (Alternatively you may wish to start the discussion with the worst animal and move up to the animals in moderate and best condition). If some households have more than one working animal, rank the animals within each household as well.
Step 5	While participants are deciding which animals are in a moderate or poor state, ask them about how they are making their decisions. What criteria are they using in their discussions? These may include the behaviour of their owners in caring for their animals, the facilities or resources being provided to animals, and observations relating to the animals themselves, such as wounds, injuries or body condition.
Step 6	Using symbols, list the criteria used to categorize each animal near to the animal or household. For example, if an animal is described as being in poor condition due to beating, poor quality of feed and insufficient space in its shelter, put symbols for each of these next to the household.
For an animal disease map use these three alternative steps instead:	
Step 4'	Ask participants if there are any animals suffering from sickness or disease right now, at this moment. Encourage them to show all the sick animals on the map, using a different symbol for each type of disease.
Step 5'	Next ask them to show past cases of disease in the animals belonging to each household.
Step 6'	Discuss the reasons for the diseases shown, when they occur (seasonal, or related to work patterns?), how they are recognized, possible sources of infection or contamination, how they spread and any other disease-related topics which the participants bring up.

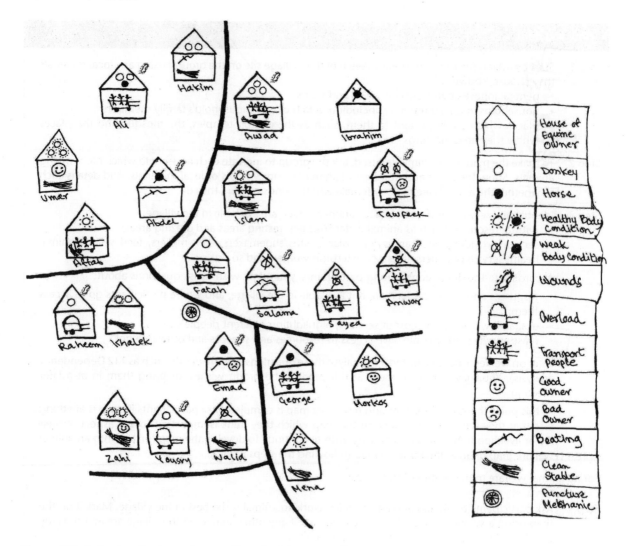

Figure T1b Animal welfare map from a village in Saharanpur district, Uttar Pradesh, India (2006)

This map shows specific animal management and handling practices (see legend), as well as the welfare status of the animals in each household. Once the households with either a horse (black dot) or donkey (white dot) were identified, the group added information about whether the owner was beating or overloading the animal and keeping the stable clean. Then they added some indicators of animal welfare: whether animals were healthy or weak and whether or not they had wounds. Based on the map, the group identified who was a 'good' and who was a 'bad' owner, exploring reasons for the management and handling practices shown, and discussing how wounds on the animal could be prevented. This led to immediate action by some of the 'bad' owners to improve their animal management and handling.

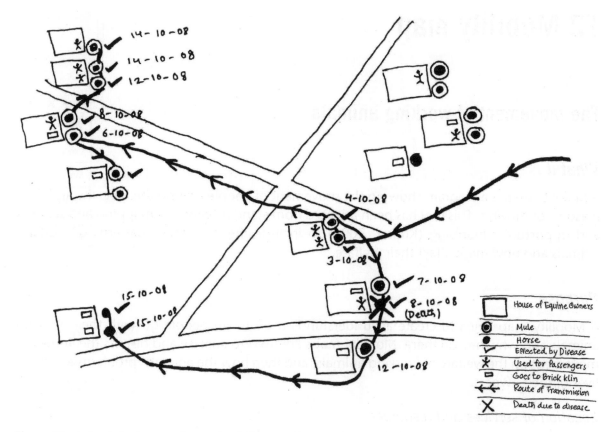

Figure T1c Animal disease map for equine influenza, Meerut, Uttar Pradesh, India (2008)

This Map was made by a group of mule- and horse-owners in a village where there was an epidemic of equine influenza. First the owners mapped their households and indicated the type of work done by each animal (see legend). Then they showed which animals were affected by influenza and when, also indicating which animals had died. Using the dates of illness the group mapped the source and route of transmission of influenza using arrows. Based on this, the owners developed an action plan for treatment of affected animals and prevention of further disease transmission.

Facilitator's Notes: Mapping

- Mapping on the ground is easiest for communities; everyone can walk around the map and see it from different angles. This also allows a large crowd to view the map and contribute to it.
- Different participants may draw different maps of the same area and that is ok. It reflects their different views of the community and of the topic discussed.
- Before starting this exercise, discuss how the map will be used. This will make the participants comfortable and more likely to share information freely
- Remember that you are not controlling the map. Give confidence to people so that they gradually take over the entire process themselves. Encourage those who are not participating.
- Intervene only when it is necessary to involve people who are left out, for example women who are watching the exercise. Ask them whether their houses, animals and other important places in the locality are shown. Ask about the various symbols used and the significance of each item to them. Understand the importance of their viewpoint.
- Discuss the map with different people while doing an Animal welfare transect walk (T22) later on, to see if it represented the ground reality.
- Maps can show how things looked in the past, in the present and what people would like their community to look like in the future. Maps can also be drawn before and after an intervention to compare changes in animal welfare or the resources and services available to working animals and their owners.

T2 Mobility map

The movement of working animals

What it is

A Mobility map is a diagram showing the movement of people around their locality and their reasons for moving. This tool has been adapted from standard mobility mapping and services and opportunity mapping (Kumar, 2002) to include the movement patterns of working animals and how these affect their welfare.

Purpose

A Mobility map helps to create an understanding of where people go with their working animals and why they go there. Mobility maps are used to analyse and discuss the impact of movement on the welfare of working animals, and then how the animals' problems affect the lives of their owners.

Location of services and resources

Mobility maps (see Figure T2 below) usually focus on the location of services and resources. This includes the movement of animals to and from their places of work and the distances travelled while working. It can also capture the distances to various services and resources, the frequency of visits and the time required for a visit; for example to get veterinary help, repair harness, collect animal feeds or take animals to water or grazing.

Opportunity map

A Mobility map can be extended into an Opportunity map, which includes the community's perception of the service providers and resources available and exploration of potential resources which are present in the area but not used. For each resource and service the Opportunity map can include its importance, cost, quality, availability and accessibility. For service providers, the map can look at the quality of service provided to both the owner and the animal.

How you do it	
Step 1	Individuals and small groups may draw maps for themselves or for 'typical' people from their village. First ask participants to list all of the destinations that they go to with their animals, including to places of work, to get farriery services, buy harness and collect animal feed or fodder. Prepare a sketch map on the ground or on chart paper with the village at the centre.
Step 2	Work with participants one at a time. Add their destinations to the map using a different symbol for each person. Ask them to talk about the frequency and reason for travel, the distance and the time spent travelling. Draw lines from their home or the village centre to each destination, using different colours or types of line (for example, dotted lines) to indicate the different reasons for travelling. Different line thickness can represent the frequency with which they travel there; for example a thick line meaning regularly and a thin line meaning infrequently. The distance may be represented by the length of line or written next to the line. Travelling time could be represented using pebbles or seeds.
Step 3	To extend the Mobility map into an Opportunity map, facilitate discussion about the importance, quality, availability and accessibility of each service provider and resource to animals and owners. Represent these on the map using pebbles, seeds and/or leaves. For example importance can be scored using pebbles and availability and accessibility can be shown using different tree leaves.
Step 4	Encourage the group to analyse what is shown on the mobility maps. Look for relationships between movement and the welfare of their working animals. Discuss the factors contributing to poor welfare and any opportunities for improvement.

Facilitator's Notes: Mobility map

- In the animal welfare context, we find it interesting to talk about the quality of the road or path. This may affect an animal's ability to pull a cart or carry a load, and the types of injury or other welfare problems that it encounters.
- The quality of a service provider may be seen from the working animal's point of view, by discussing how he or she treats the animal. For example: Is the hoof-trimmer gentle, not frightening the animal and not causing pain?

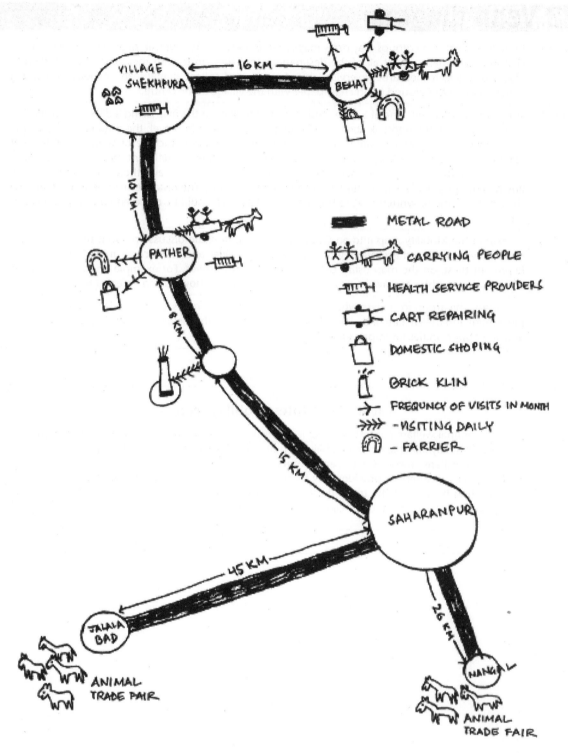

Figure T2 Mobility map showing animal-related services and resources, Shekhpura village, Saharanpur district, Uttar Pradesh, India (2008)

The Mobility map on page 152 (Figure T2) was drawn by a group of mule owners. It shows that most movements are made to and from Behat and Pather villages, to transport people, get carts repaired, do domestic shopping and use farriery and animal health services. It also shows travel to brick kilns during their working season and to trade fairs at Nanga and Jalalabad for buying and selling animals. The thick line represents a metalled road. Thin lines with arrows represent the frequency of visits.

T3 Venn diagram

Animal-related service providers, animal diseases

What it is

A Venn diagram uses circles of different sizes to represent relationships. In the original form it represents relationships between the community and particular individuals or institutions. We have adapted it for analysis of relationships between animal owners and animal-related service providers. We have also extended it to look at the impact of disease on working animals.

Purpose

Service provider Venn diagram

This tool (see Figure T3) enables community members to identify all the animal-related service providers in their locality, including farriers or hoof trimmers, feed sellers and animal health providers. It can help participants to analyse their relationships with service providers and the usefulness, availability, accessibility, cost, quality and importance of the service provider to their own life or to the welfare of their working animals.

Animal disease Venn diagram

This adapted tool can be used to analyse the animal disease situation in a village. The animal (rather than the owner or community) is the focal point for analysis.

	How you do it
Step 1	**Service provider Venn diagram** Start by asking participants to list all the stakeholders or service providers who are relevant to them and their working animals. Identify the **most important** service provider in terms of his or her contribution to the working animal or to the participants' own lives, then the next most important, and so on. Ask participants to draw an animal (or a symbol for the village or group) in the centre. Represent each service provider with a circle made from paper, stones or chalk, using different sizes of circle to show their relative importance – the biggest circle for the most important and the smallest circle for the least important. Ask participants to place the stakeholder circles around the animal at the centre. Indicate the name of the stakeholder on each of the circles.
Step 2	Ask the group which service providers they have good relationships with and why. Move the service provider circles, putting them closer or further away from the centre according to whether the relationship is better or worse. We find that there is often a lot of debate and discussion while positioning the circles. An alternative is to use the placing of the stakeholder circles to represent their accessibility or availability. This can be developed using beans, seeds or stones to add more dimensions; for example scoring the cost or quality of an animal health provider, or the frequency of use.
Step 3	When the Venn diagram is complete, encourage participants to discuss what it shows. Ask questions such as: • How do you feel about the stakeholders' role in your community? How do you find them helpful? • In what ways are you satisfied or dissatisfied with service providers in the community? • What could they do to serve you and your working animals better? How could you help them to improve their services to the community?
Step 4	**Animal disease Venn diagram** Follow the steps above, but use the size of each circle to represent the relative importance of each disease affecting working animals, in terms of the severity of its effect on the animal itself. The distance between the centre and each disease circle represents the frequency of disease occurrence. Beans or seeds can be used to score the cost of treatment or prevention, and the effects of animal disease on the owner and his or her family. Discussion points may include: • How does each disease affect the animal? What are the signs? • How does each disease affect the owner and his or her family? How does it affect the community as a whole? • What can be done to treat the disease? • What can be done to prevent the disease?

Facilitator's notes: Venn diagram

• This method is often confused with Mobility mapping (T2).
• If using paper circles, you can cut out a selection of different sizes ahead of time. Using different coloured circles provides a nice contrast.
• Local resources may also be used, such as cut banana leaves.
• Circles can be drawn on the ground with coloured dust, chalks or using a wooden stick.
• Seeds poured onto a plate can also be used to make the circles. People can make divisions in the seeds to show proportions. If they change their minds they can easily change the size of each division in the seeds.
• You can examine the influences and quality of services with different groups in a community – such as men, women and children – to get a better insight into the perception of the whole community.

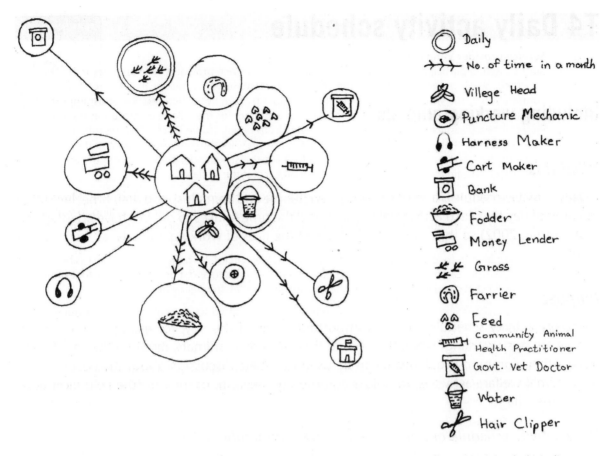

Figure T3 Venn diagram of animal-related service providers and resources in a village, Uttar Pradesh, India (2006)

The centre circle represents the village. Outer circles show the relative importance (according to size) of the services and resources used by animal owners. In this diagram, circles are placed at different distances from the centre according to their accessibility; for example the government veterinary hospital has been shown in a small circle and far away, because animal owners feel that the service is of low importance to them and often the vet is not available. Arrow-heads represent the frequency with which people use each service.

T4 Daily activity schedule

Including working animals

What it is

A Daily activity schedule is a chart showing how people and animals spend their time, including the time of day that each activity takes place and the time it takes to do. We have adapted this tool (Kumar, 2002) so that it includes aspects of the lives of working animals, as well as their owners.

Purpose

The Daily activity schedule identifies important times of the day, for example times when people are busy working, when they spend time with their animals, or when they are free to discuss their common problems. It can be used to initiate discussion about the best times to plan animal welfare activities, to hold a community meeting, or for you (the field facilitator) to visit the village.

Daily activity schedule of the animal-owning community

This chart (Figure T4a) explores and compares how animal owners, users, handlers and carers spend their work and leisure time. It helps people to understand the roles and responsibilities of different members of the family towards their animals. It can be used to analyse the factors that influence different people's roles and activities, and to understand the problems and obstacles faced when dealing with animals.

Daily activity schedule of the animal

This chart (Figure T4b) explores how working animals spend their day, during both work periods and rest periods. It can be used to look at a daily routine from the animals' point of view and find out where improvements to welfare could be made, such as increasing the time available to animals for rest or grazing.

How you do it	
Step 1	Explain the exercise to the participants. Daily activity charts are best made by individuals and small groups, so divide up larger groups to make charts for different people, such as men, women and children.
Step 2	Start a discussion about the activities that a person, a group of people or their animals normally do, from when they get up in the morning until they go to sleep in the evening. Ask them to list all the activities in order.
Step 3	Agree whose daily activities to chart first. This can be the animal owner, members of the owner's family, a single animal or a group of animals. Agree whether to make a circular clock or a line chart to represent time. Decide whether to show time in hours or as parts of the day, such as morning, afternoon and evening. Show daily activities using symbols placed at the appropriate time of day.
Step 4	Discuss the Daily activity schedule with participants. Key points can include: • free time and working time • periods of heavy workload or strain • times when animals are fed or the animal shed is cleaned • when are animals taken for grazing? • when are animals offered feed and water? • when do people groom animals or clean their hooves? • how much the activity is liked or disliked? • how participants feel about the way they use their time throughout the day? Also ask when people have free time to take part in other activities, either individually or as a group.

Facilitator's notes: Daily activity schedule

- Daily activity charts are a useful, non-threatening, rapport-building exercise to look at real life experiences.
- You might want to discuss how daily routines change depending on the time of the week or the season.

DAILY ROUTINE ACTIVITIES
(Men, Women and Children)

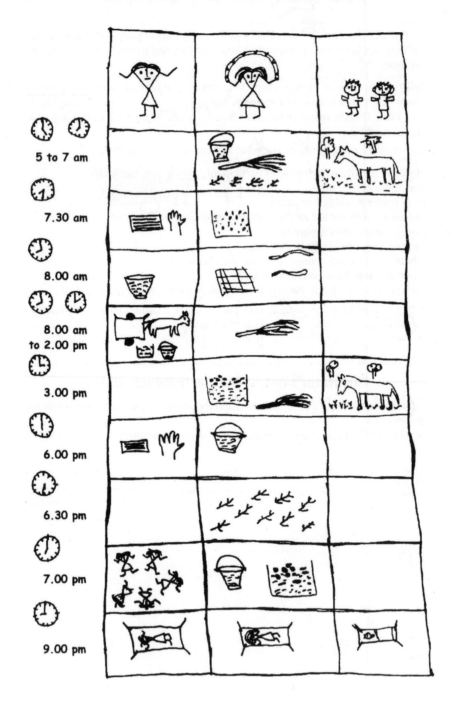

Figure T4a Daily activity schedule of an animal-owning community in Ghaziabad, Uttar Pradesh, India (2008)

In this exercise (Figure T4a above), men, women and children listed all their activities between getting up at 5 am and going to bed at 9 pm. Women start cleaning the animal shed and watering their animals between 5 and 7am. Men feed the animal around 8am and then take it out to work. Women are involved in feeding, watering and cleaning for animals several times a day, while men are involved with their animal only between 8am and 2pm. Children take animals for grazing between 3 and 6pm. This initiated a discussion about the roles and responsibilities of family members in animal care.

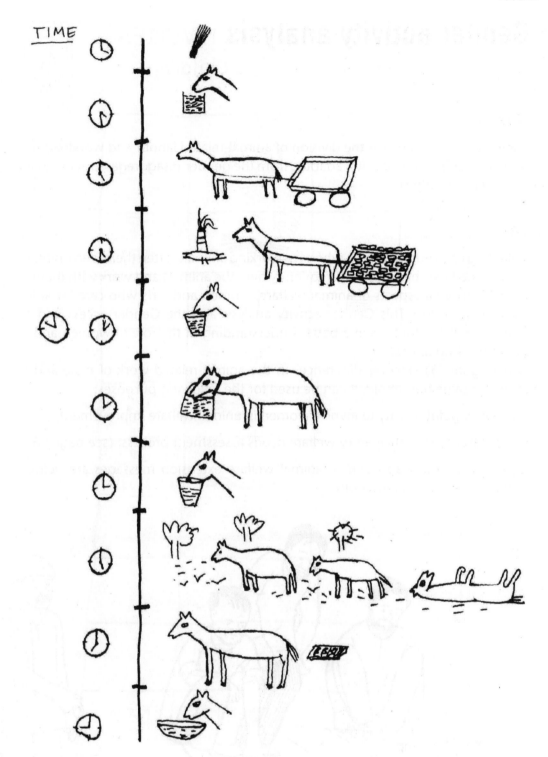

Figure T4b Daily activity schedule of the working animal in Meerut, Uttar Pradesh, India (2008)

This Daily activity schedule (Figure T4b, above) for animals was facilitated with brick kiln workers, in order to initiate a discussion on animal welfare issues. The animals are used for transporting bricks by cart. The day starts with cleaning the feeding trough at 4 a.m., followed by feeding at 4.30 a.m. At 5 a.m. the animal is harnessed to the cart and work starts at the brick kiln at 5.30 a.m. Water is offered between 10 a.m. and midday and again at 3 p.m. when the animals return home. A second feed is offered after finishing work at 2 p.m. From 5 to 6 p.m. animals can graze and roll. They are groomed between 7 and 8 p.m. and the last feed is given between 8 and 9 p.m.

T5 Gender activity analysis

What it is

Gender activity analysis explores the division of animal-related labour and workload between men and women, boys and girls. It is adapted from the gender-disaggregated activity calendar (Thomas-Layter et al, 1995).

Purpose

While facilitating community groups who own working animals, often there is an unintentional focus on men and boys, because they frequently own the animals and work with them during the day. To improve all aspects of animal welfare, women and girls who care for animals at home must be included. This Gender activity analysis and the Gender access and control (T10) tools are both useful to gain a better understanding of the role of women and girls in looking after working animals.

This tool (Figure T5) explores differences in the animal-related work of male and female members in the same household. It can be used for the following purposes:

- as an entry point activity to involve women in animal welfare improvement;

- as the start of the participatory welfare needs assessment process; (see page 86)

- to ensure the most appropriate animal welfare extension messages are delivered to each member of the household.

How you do it	
Step 1	Explain the exercise to the participants. Gender activity analysis is better when carried out in small groups. We find it works best in separate gender groups first, with each group analysing the work for both male and female members of the household. Then the two groups may be combined for further discussion.
Step 2	Start by asking about the activities that men and women normally do when looking after their working animals. Include activities carried out on a daily, weekly or monthly basis. Ask the group to list these on the ground or on a chart, using symbols or drawings of the activities.
Step 3	Ask the group to indicate how the workload for each activity is divided or shared between men and women, using up to ten seeds or stones to score their relative contributions (see Figure T5).
Step 4	Discuss the gender activity analysis with the two groups separately. Then bring the two groups together to discuss any differences between the men's and women's chart. Key points for discussion can include: • Who is responsible for which activity and why? • Which activities are done when the animal is at home and which are done when the animal is at work? • Which activities are typically done by men and women, and why? • How much time does each activity require and is the workload (number of activities) equally distributed? • Which animal-related activities are done by children and why? Is there a difference between boys' and girls' work? • What are the implications on the welfare of the animal, if any, if a man, woman or child does the activity? • Would participants like to change anything in the division of labour shown in their diagram? If so, what and why?

Facilitator's notes: Gender activity analysis

• Looking after the animals in the household is usually the responsibility of all family members, although different people are responsible for different tasks.
• Girls and boys can also do this analysis, by adding an additional two columns to the example shown in Figure T5.

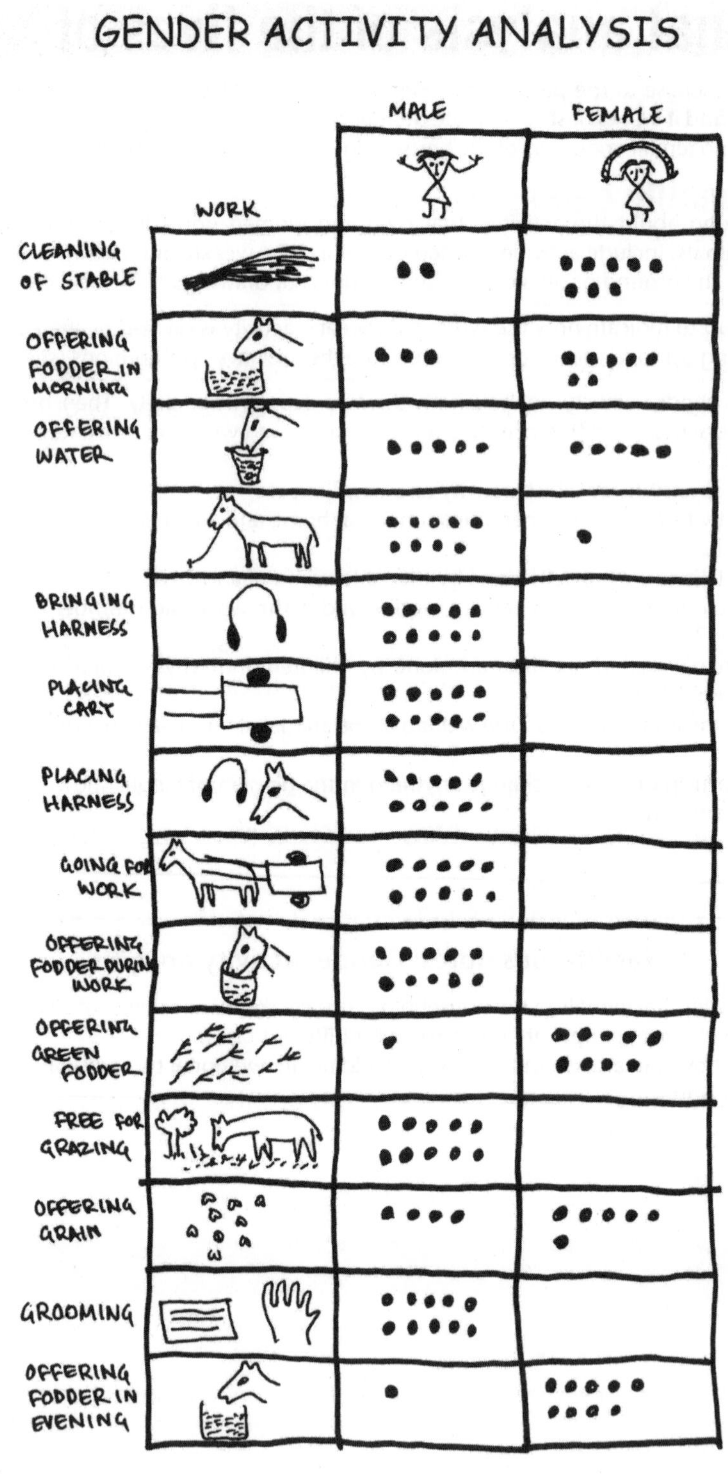

GENDER ACTIVITY ANALYSIS

WORK		MALE	FEMALE
CLEANING OF STABLE		••	••••• •••
OFFERING FODDER IN MORNING		•••	••••• ••
OFFERING WATER		•••••	•••••
		••••• •••••	•
BRINGING HARNESS		••••• •••••	
PLACING CART		••••• •••••	
PLACING HARNESS		••••• •••••	
GOING FOR WORK		••••• •••••	
OFFERING FODDER DURING WORK		••••• •••••	
OFFERING GREEN FODDER		•	••••• ••••
FREE FOR GRAZING		••••• •••••	
OFFERING GRAIN		••••	••••• •
GROOMING		••••• •••••	
OFFERING FODDER IN EVENING		•	••••• ••••

Figure T5 Gender activity analysis for work with animals, Ghaziabad, Uttar Pradesh, India (2008)

The chart (Figure T5) explains the animal-related work distribution of men and women in a village in Ghaziabad, Uttar Pradesh, India. The group used 10 pebbles to score the division of labour for each of fourteen daily activities. Women have more responsibility for cleaning the animal shelter, offering fodder and grain and watering animals. Men do all the activities related to working the animals, harnessing, taking animals to graze and tying them up. Participants discussed why some of the activities were only carried out by men or women and what effect this had on their animals.

T6 Seasonal analysis of the lives of working animals

Disease seasonality calendar

What it is

A Seasonal analysis (also called a Seasonal calendar, Seasonal diagram or Seasonal activity profile) is a diagram of changes over the annual cycle of months or seasons. We have adapted this tool from a similar tool (Kumar, 2002) so that it includes aspects of the lives of working animals as well as their owners.

Purpose

This tool enables people to analyse how their livelihoods and the welfare of their working animals change in different seasons, and how these changes influence each other. For example, it may highlight changes in animal welfare status according to seasonal changes in the workload or types of animal feed available. It helps the community to decide on actions to improve animal welfare and plan ahead to prevent welfare from getting worse in a difficult season.

Seasonal analysis of the lives of working animals and their owners

The analysis (see Figure T6a) can cover many aspects of life, using scoring to show the size of seasonal variations in:

- livelihood activities of animal owners, users and carers as different work is done at different times of year;
- availability of work or employment;
- migration patterns of animal-owning families;
- work load of animals, periods of heavy work and periods of relative ease;
- availability of animal feed and fodder, grazing or other resources;
- animal diseases;
- changing welfare status of animals.

How you do it	
Step 1	The Seasonal calendar is best carried out in small or medium-sized groups, so divide larger groups to compare how seasonal changes affect different people.
	Discuss the local calendar and seasonal landmarks; for example months, dry or rainy seasons, festivals and religious ceremonies. Ask participants to identify unique characteristics of each month or season and depict these using a symbol or drawing.
Step 2	Start the discussion with the present season and the work that that participants do during this season. Agree on which activities, events or problems are going to be discussed and mark changes in these along the calendar line. Participants might find it easier to start by discussing general issues such as the types of work and income flow at different times of the year, before moving on to animal-related issues such as diseases or the availability of feed, water and grazing.
Step 3	Encourage participants to score seasonal changes using beans or seeds. For example, higher income levels can be shown by placing a lot of seeds on the month in which this occurs, while a decrease in income the next month is shown with just a few seeds.
Step 4	Discuss reasons for the seasonal changes shown. Explore the relationships between different seasonal events, activities and problems for working animals. Discuss and agree action points for improvement of animal welfare or preparation for difficult seasons.

Disease seasonality calendar

The seasonal calendar (Figure T6b) can be made specific to one area of concern, such as a specific pattern of animal disease.

MONTHS	चैत्र MARCH - APRIL	बैसाख APRIL MAY	जेठ MAY JUNE	आषाढ JUNE JULY	सावन JULY AUGUST	भादो AUGUST SEPT.	आसिन SEPT. OCT.	कार्तिक OCT. NOV.	अगहन NOV. DEC.	पूस DEC JAN	माघ JAN FEB.	फागुन FEB MARCH
DISEASE	••• •••	•••• ••••	•••• ••••	••• •••	•• ••	•• ••	•			••	••	•
INCOME	•••• ••••	•••• ••••	•••• ••••	••	••	••	•••	•••	•••	•••	•••	•••
INCOME FROM OTHER SOURCES	••••	•••	•••	••	••	•			••	•		•• ••
EXPENDITURE	•••• •••	•••• ••••	•••• ••••	••	••	••	•••	•••	•••	•••	•••	•••
FODDER	⌂ ⌂	⌂	⌂	⌄⌄	⌄⌄	⌄⌄	⌂⌂	⌂	⌂	⌂	⌂	⌂
ANIMAL'S BODY CONDITION	••	••	•	•	•	•	••• •••	••• •••	••• •••	••• •••	••• •••	••• •••

Figure T6a Seasonal analysis of the lives of working animals and their owners, India (2008)

Seasonal analysis (Figure T6a above) was one of the first exercises carried out with a community in Hardoi, Uttar Pradesh, India as part of rapport building, described in Chapter 4. Working with animals is the main source of village income (second row) and the group has two additional sources of income: wage labour and sale of agriculture produce. Overall income is lowest in the summer months. Green fodder is available during rainy months (grass symbol) and in other months the animals receive dry fodder (grain symbol). Relationships between the body condition of animals and seasonality of feed and disease were analysed by the group: animals are in good condition between September and November because there is more feed available and less risk of disease than in other months.

Facilitator's notes: Seasonal calendar

• It is important to enable people to use their own way of measuring time if a twelve-month calendar is not used. Instead of scoring with beans or seeds, sticks of different length can be placed along the calendar to show increases and decreases over time, or both seeds and sticks can be used for different activities.

SEASON / मौसम	WINTER जाड़ा	SUMMER गर्मी	MONSOON बरसात
Respiratory Problem / धसका नजला	● ● ● ● ● ● ●	●	● ●
Surra / सुर्रा	● ●	● ● ● ● ●	● ● ● ●
Colic / पेट दर्द	● ● ● ● ● ●	● ●	● ●
Tetanus / ट्रिशणब्याल	● ● ●	● ● ●	● ● ● ●
Lameness / लँगड़ापन	● ● ● ● ● ●	● ●	● ●
Hoof Problem / खुराली	● ● ●	●	● ● ● ● ● ●
Eye Problem / आँख में पानी आना	● ●	● ● ● ● ● ●	●
Foot Cankar / चकरावल	● ● ● ● ● ●	● ●	● ●
Hoof Swellings / सुम में छाले	● ●	● ● ● ● ●	●
Loose Motions / दस्त	● ● ● ●	● ● ● ●	● ●
Wounds / जख्म	●	● ● ● ●	● ● ● ● ●

Figure T6b Disease seasonality calendar, India (2008)

A community group in Ghaziabad, Uttar Pradesh, India carried out a seasonal analysis of diseases affecting their donkeys, as part of developing a shared vision and collective perspective (see Chapter 4). Eleven diseases were identified and scored against three different seasons: winter, summer and rainy season. Respiratory problems, colic, lameness and foot canker occurred most often in winter, while surra, eye problems and hoof swelling were more common in summer. In the rainy season, wounds and hoof problems are the biggest disease issues. Based on this analysis the group discussed why some diseases were seasonal and how they could be prevented.

T7 Historical timeline

What it is

A Historical timeline lists the past events in a community in chronological order. This tool works particularly well when carried out with the elderly people of the village. It is a good rapport building exercise with the community.

Purpose

A Historical timeline exercise (figure T7) will provide both you and the participants with an insight into how each person perceives past history and which events are seen as important. In the animal welfare context it is interesting to draw a timeline of the history of working animals in the village, including important events such as when the first animal was brought into the village, outbreaks of disease, and introduction of the first mechanized transport into the area.

	How you do it
Step 1	Start the discussion on past events using questions such as: Was there ever a time when there were no working animals in the village? When was that? When did the first one arrive?
Step 2	As participants recall events, ask a group member to write them on cards or show them using symbols on the ground. Ask if people can remember these events in a specific year or related to any other important or well-known event in the country. This will enable you to work out the year. Discuss and ask questions about as many events as they can remember that relate to the history of working animals in their locality.
Step 3	Then ask the group to organize all the events in time order, starting with the earliest events at the top and gradually adding all later events below, until they reach the most recent event at the bottom. When they have finished the timeline, ask them to check if all the events are there and the timeline is correct.
Step 4	Discuss thoroughly the aspects and events that participants are interested in. Bring up events for discussion that you are particularly interested in and explore more of their history with the group.

Facilitator's notes: Historical timeline

- Initially participants may find it difficult to relate events to the time when they happened. We often find that people use a local time frame which is different from our calendar years. You will need to use your own judgement to enable people to articulate time in a way that both you and they can relate to.
- Recalling or remembering all the events in a timeline is not the main purpose of the exercise. The points discussed during the process of drawing the timeline are more important than the time line itself, which captures very little information. Important animal-related issues which arise while the group is making the timeline should be taken forward for further exploration by the group.

HISTORICAL TIME LINE

1938		FIRST TEMPLE CONSTRUCTED
1939		LATE KAMLESH CHANDRA PASSED HIGH SCHOOL
1945		FIRST PRIMARY SCHOOL STARTED
1952		DURGA PRASAD ELECTED AS VILLAGE HEAD
1960	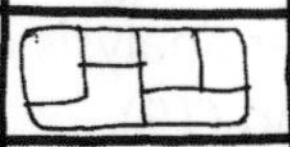	FIRST LAND ALLOTMENT
1965		FIRST ROAD CONSTRUCTED
1981		FIRST INDIA MARK HAND PUMP INSTALLED
1985		FIRST BRICK KLIN ESTABLISHED
1986		ELECTRICITY IN VILLAGE
1988		FIRST BICYCLE OF THE VILLAGE PURCHASED BY RAMSINGH
1990		FIRST HORSE PURCHASED BY TEJPAL
1991	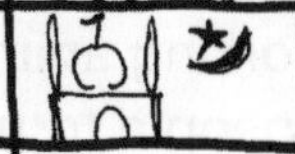	FIRST MOSQUE CONSTRUCTED
1992		FIRST TELEPHONE
1994		FIRST MOTOR BIKE PURCHASED BY KALLU
1996		METALLED ROAD CONSTRUCTED
1998		SAHEB LAL BOUGHT SECOND HORSE OF THE VILLAGE

Figure T7 A historical timeline from Sunni village, Hardoi district, Uttar Pradesh, India (2009) including key animal-related events

The Historical timeline in Figure T7 begins in 1938 when the first temple was built in Sunni village, Hardoi district, Uttar Pradesh. The first horse was purchased by Tejpal in the village in 1990 and the second horse by Sahib Lal in 1998. This exercise was done as part of the initial interaction with the community during Phase 1, Feeling the pulse (Chapter 4) and generated an interesting discussion on horse ownership and use in the village.

T8 Pair-wise ranking and scoring

What it is

This tool uses a matrix to make direct comparisons between items or issues, without referring to a scoring system or any other pre-determined criteria. Adapted from a similar method (Kumar 2002), it helps prioritise different issues for further analysis or action. It is often used before a more detailed ranking tool, such as Matrix ranking (T9).

Purpose

Pair-wise ranking helps people to compare and prioritise things, such as different service providers, animal diseases or varieties of animal feed, in order to arrive at a decision about which one the community prefers. It is also used to prioritise welfare issues for taking action, after a participatory welfare needs assessment has been carried out (see Chapter 4, Phase 3). Several examples of pair-wise ranking follow that we often use with animal owners.

Pair-wise ranking of animal diseases

The chart (Figure T8a) compares common diseases in working animals and identifies which of them are perceived to be a bigger problem. It helps people to understand the common diseases in their location and can be used as the basis for the Animal welfare cause and effect analysis tool (T26).

Pair-wise ranking of animal-related service providers

Pair-wise ranking (see Figure T8b) may be used to analyse several aspects of service providers, such as the importance of the service provider for the welfare of animals, ranking of the quality, cost or availability of different service providers (for example the animal health worker, farrier, saddler and feed seller) or ranking several providers of the same service, such as the farriers in one locality.

   A second dimension of analysis may be added to the matrix as shown in Figure T8b below. As well as ranking the importance of various service providers to animal welfare, the group also used arrows on the same chart to indicate which service provider they preferred over the others in terms of service quality and friendliness.

Pair-wise ranking of animal welfare issues

This helps the community to prioritise the welfare issues identified during the Participatory welfare needs assessment.

How you do it	
Step 1	Discuss the subject to be explored, for example which diseases of working animals are seen to be a severe problem for the village. Ask participants to discuss the most common diseases amongst themselves and come up with a detailed list, making sure that no diseases are missed out.
	Use symbols to depict the diseases, such as different tree leaves, coloured cards or other locally available materials. Make two sets of each symbol.
Step 2	Ask the group to draw a grid or matrix on the ground using coloured powder, chalk or a stick.
	Place one set of symbols in a column, from top to bottom. Then place the second set of symbols in a line from left to right (or right to left if the group prefers), in the same order. One by one, compare each symbol in the columns with each symbol in the rows. Encourage participants to discuss which one of the pair they perceive to be a bigger problem and why. Put the preferred symbol in the relevant box on the matrix. Cross out the boxes in which same symbols are compared. This will lead to the bottom half of the boxes being crossed out because they repeat what is on the top (see Figure T8a).
Step 3	Count the number of times each option appears in the matrix. Add them up and record totals at the bottom of the matrix using seeds or pebbles. Then make a list of the options with the most-preferred option (or the most severe disease problem) ranked first.
	Discuss the results of the activity. The next step could be to carry out a root cause analysis of the disease perceived as the biggest problem, or a decision to invite the most-preferred service providers to a group meeting in order to plan with them for action to improve animal welfare.

Facilitator's notes: Pair-wise ranking

- If the number of items to be compared is too large, this exercise can become boring for participants. In this case reduce the list by focusing on a smaller number of items.
- The discussions that people have about why they choose one option over another are just as important as the result. These reasons for choosing one option over another should be recorded.

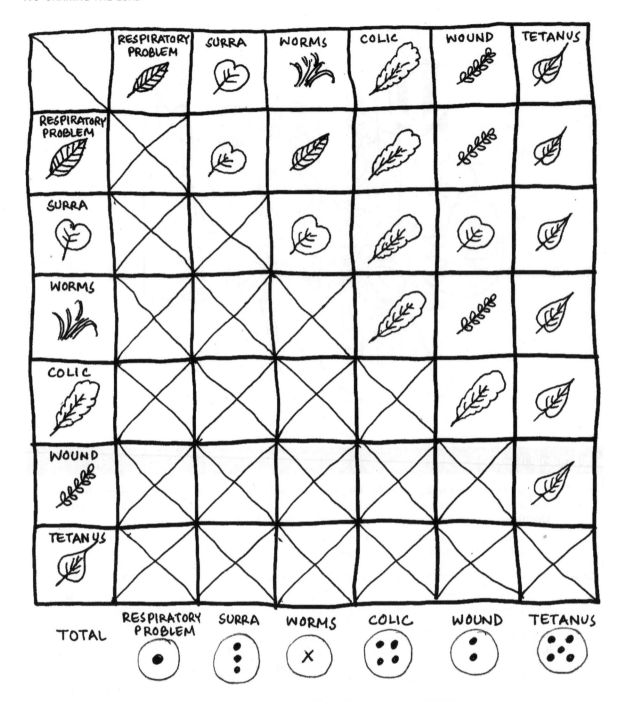

Figure T8a Pair-wise ranking of animal diseases, Meerut, Uttar Pradesh, India (2007)

A group of horse owners in Meerut, Uttar Pradesh, identified, compared and ranked six diseases, using tree leaves as symbols for respiratory problems, surra (*trypanosomiasis*), worm infestation, colic, wounds and tetanus. This showed that tetanus was seen by the group to be the biggest welfare problem, followed by colic. Though worm infestation was initially identified as one of the most important diseases, it was found to be less important than other diseases during pair-wise comparisons. This tool was used as part of the situational analysis described in Chapter 4, Phase 2. As a result, the group decided to take up community-led vaccination of their animals against tetanus (see Cost-benefit analysis, T15).

PAIR WISE RANKING (SERVICE PROVIDERS)

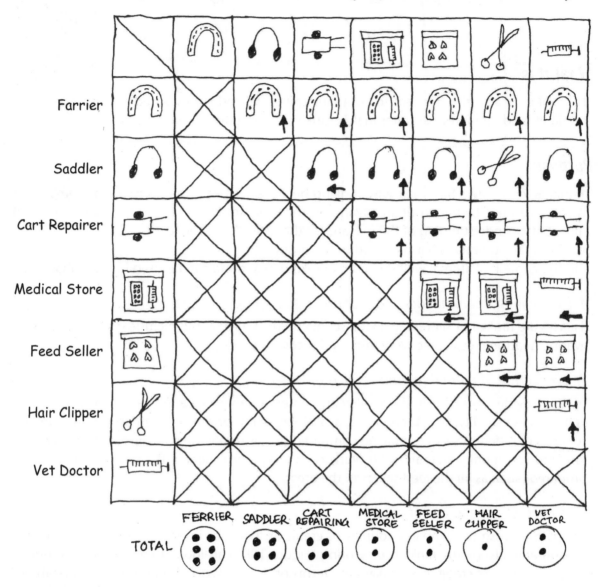

Figure T8b Pair-wise ranking of service providers, Abupur village, Saharanpur, Uttar Pradesh, India (2007)

A group of animal owners in Abupur village, Saharanpur, Uttar Pradesh ranked service providers in their locality according to who they found most important for the welfare of their animal. The farrier was seen as the most important. The group also included a second dimension in the matrix, using arrows to indicate which service providers they preferred in terms of quality and friendliness.

T9 Matrix ranking and scoring

What it is

This tool uses a matrix diagram to compare animal-related issues based on pre-determined criteria. It is adapted from another matrix ranking/scoring tool (Kumar, 2002).

Purpose

Matrix ranking differs from pair-wise ranking because it ranks or scores issues or items based on criteria agreed by the group in advance of the exercise. Both you and the participants will gain a better understanding of the reasons for the group's preferences or choice and how the decision-making process happens within the group:

In the context of animal welfare, matrix ranking or scoring can be used for the following:

Matrix ranking of animal diseases (or other animal welfare issues)

Diseases are ranked or scored against a set of criteria, which could include disease frequency, severity, treatment costs, recovery rates and recovery times.

Matrix scoring of animal-related service providers

Service providers available in the locality are scored or ranked (see Figure T9a) against criteria such as service quality, cost, friendliness, availability or importance for working animal welfare. Providers of different services may be compared (for example the hair clipper, farrier, agrovet store and community animal health worker) or several providers of the same service may be compared (such as all the feed sellers in the area).

Matrix scoring of sources of credit

Groups may wish to compares sources of credit or loans for improving the welfare of their working animals. Matrix ranking or scoring (Figure T9b) can look at criteria such as interest rates, availability of credit, easy repayment terms or risk of losing property or land if unable to repay on time. Sources of credit may also be identified through the Credit analysis tool (T13).

How you do it	
Step 1	Identify the issues or items to be analysed for decision-making, based on previous discussions or exercises with the group on the agreed topic. Discuss the reasons or criteria that will be used for making decisions about which items will be preferred over others. If a lot of reasons or criteria are given, encourage the participants to sort out which are the most important ones. Make a list of these criteria. Criteria should be either all positive or all negative: mixing positive and negative criteria will create confusion later.
Step 2	Draw a matrix on the ground with the criteria listed from top to bottom and the items for ranking listed from left to right. Ask the group to rank all the items based on the first criterion they have chosen. Then rank them all based on the second criterion and so on, until the full list of items has been compared against all of the agreed criteria.
Step 3	When the ranking is complete, facilitate the group to draw conclusions from the exercise by asking questions. For example, ask about the items which ranked first and last, encouraging a more in-depth discussion of the reasons for these decisions.
	Matrix scoring is done in the same way but using up to ten seeds or stones to score each criterion instead of ranking them. Allow enough time for in-depth discussion and analysis of the reasons for ranking or scoring and enough time to come to consensus. The entire matrix should be documented for further decision-making and action planning.

Facilitator's notes: Matrix ranking and matrix scoring

- Consciously or unconsciously, there is often a tendency amongst facilitators to include their own criteria for ranking or scoring, rather than those of the community. It is important that you enable and allow the community to come up with their own criteria. Some of the criteria selected by the community may look strange, but if participants consider them to be important, this needs to be respected and their rationale understood.
- We have used these tools effectively for monitoring and evaluating changes in items or issues over time, as our welfare improvement programme progresses.
- This is a versatile exercise and is open to improvisation and innovation.

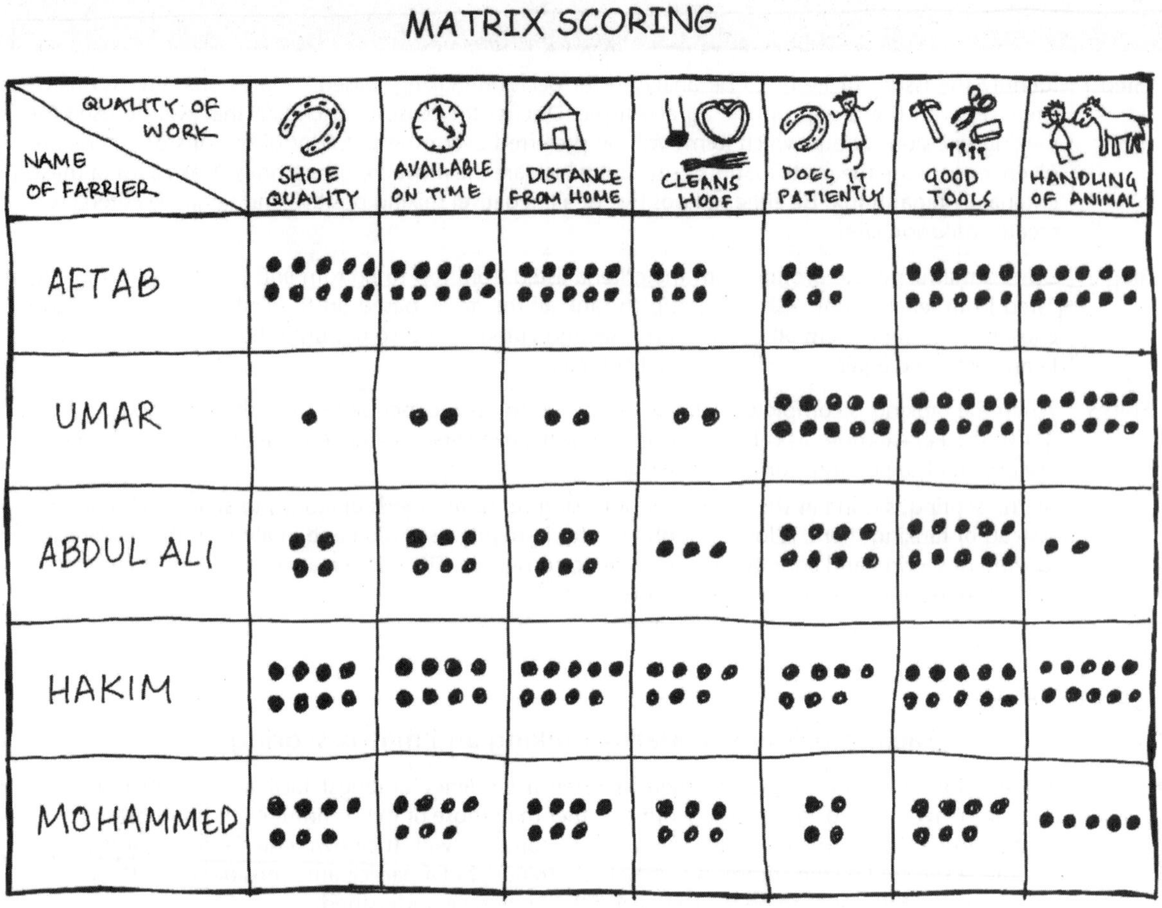

MATRIX SCORING

QUALITY OF WORK / NAME OF FARRIER	SHOE QUALITY	AVAILABLE ON TIME	DISTANCE FROM HOME	CLEANS HOOF	DOES IT PATIENTLY	GOOD TOOLS	HANDLING OF ANIMAL
AFTAB	●●●●●● ●●●●●	●●●●●● ●●●●●	●●●●● ●●●●●	●●● ●●●	●●● ●●●	●●●●● ●●●●●	●●●●● ●●●●●
UMAR	●	●●	●●	●●	●●●●● ●●●●●	●●●●● ●●●●●	●●●●● ●●●●●
ABDUL ALI	●● ●●	●●● ●●●	●●● ●●●	●●●	●●●● ●●●●	●●●●● ●●●●●	●●
HAKIM	●●●● ●●●●	●●●● ●●●●	●●●●● ●●●●	●●●● ●●●	●●●● ●●●	●●●●● ●●●●	●●●●● ●●●●
MOHAMMED	●●●● ●●●	●●●● ●●●	●●●● ●●●	●●● ●●●	●● ●●	●●●● ●●●	●●●●●

Figure T9a Matrix Scoring of the quality of service provided by farriers, India (2007)

The services of five local farriers in Saharanpur, Uttar Pradesh were compared using a Matrix scoring exercise in a horse-owning community. The community group set seven criteria, including patience and good animal handling, as important aspects of a good quality farrier. They scored each criterion using ten seeds. The group decided to invite the best two farriers to their next meeting, in order to build better relationship with them and negotiate a group rate for their services.

CREDIT ANALYSIS

SOURCES OF CREDIT	ELIGIBILITY	RATE OF INTEREST	MORTGAGE	RETURN OF INTEREST	GUARANTOR	LEGAL ACTION	AVAILABLE IN TIME	THREAT				TOTAL
								LOSS OF ORNAMENTS	LOSS OF LAND	LOSS OF GUARANTOR	LOSS OF LIFE	
MONEY LENDER	X	oo oo	(ornaments + cultivated land)	X	(guarantor)	|||||	ooo oooo ooo	X	ooo ooo ooo o	ooo oo	X	ooo ooo oo (8)
BANK	(cultivated land)	o	(cultivated land)	X	(guarantor)	|||| ||||	ooo	X	oooo oooo	ooo ooo oo	X	ooo oo (5)
CO-OPERATIVE SOCIETY	(cultivated land)	o	(cultivated land)	X	(guarantor)	|||| |||	ooo o	X	oooo oooo	oooo oooo	X	ooo oo (5)
RELATIVES	⊙	X	X	X	X	X	ooo oo	X	X	X	ooo oo	ooo oo (5)
GROUP SELF HELP	⊗	oo	X	o	X	|||	oooo oooo oo	X	X	oo	X	oooo oooo oo (10)

KEY

(cultivated land symbol) Cultivated land

(ornaments symbol) Ornaments

(guarantor symbol) Guarantor

T9b Matrix scoring of 5 different sources of credit, based on 11 criteria identified by a group of animal owners in Rupak village, Unnao, Uttar Pradesh, India (2009)

A group of animal owners from Rupak village, Unnao, Uttar Pradesh, analysed five different sources of credit against eleven agreed criteria. The exercise revealed some high levels of risk (threat) associated with borrowing money, including losing their land, guarantor or even their life if the money was not repaid. Through this exercise the group realized that the only way to benefit from the interest would be to borrow from their own self-help group. This initiated the interest and action to start saving money as a group, which could then be lent to members to help their animals and their families.

T10 Gender access and control profile

What it is

This tool analyses which members of a community have access to resources and services and how these resources and services are controlled. We have adapted it from its original (Thomas-Layter et al, 1995). This version focuses on animal-related resources such as water, feed and grazing, and on the services provided to working animals by outsiders, such as foot-trimming or veterinary treatment.

Purpose

The Gender access and control profile (see Figure T10) helps to create a common understanding of access to, and control of, animal-related resources and services by different members of a household or village. This understanding helps participants decide what type of action is needed to improve their animals' welfare, who can do it and who will benefit from a particular action. For example, we have used this tool to look at family decision-making around the use of veterinary services.

The profile is often used to explore differences in access and control between men and women in the village or household. Children may also be included. Another option is to look at differences in access and control between animal owners, animal carers and animal users or hirers.

How you do it	
Step 1	Ask the group to list all the resources and services they use to take care of their working animals. Draw a table and list all the resources and services down the side. Along the top make two wide columns, marked 'access to' and 'control over' (in figure T10 'access to' is replaced by 'who does it?'). Divide the 'access' and 'control' columns into as many sub-columns as you need - one for men and one for women, or three for 'owner', 'user' and 'carer', for example.
Step 2	Facilitate the group to score all the listed resources and services relative to each other, using seeds or pebbles. Group members can give a score out of 10 for each item (or any number that they decide) according to who has access to the resource or service and who has control over it.
Step 3	Enable participants to analyse the scores they have given and the differences between access and control by men and women. What is the effect of the difference on people and their working animals? Discuss whether changed or increased access and control over specific resources and services would benefit animal welfare and if so, what action could be taken to achieve this and who could take it.

Facilitator's notes: Gender access and control profile

- Before doing this exercise, decide whether it would be best carried out with men and women (or other subgroups) separately or in a mixed group. This will depend on the prevailing situation and your rapport with the community.

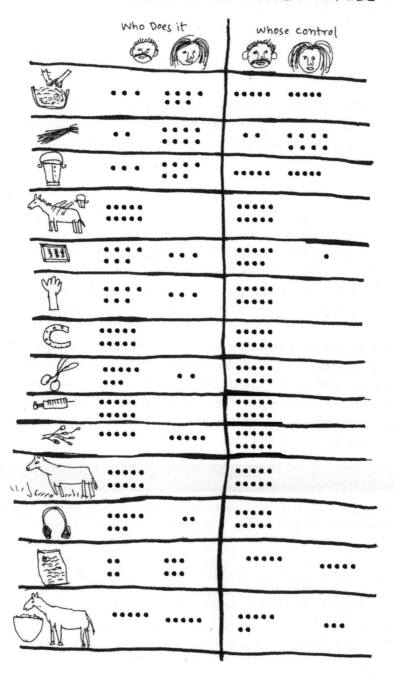

GENDER ACCESS & CONTROL PROFILE

Figure T10 Gender access and control profile of animal welfare services and resources, Saharanpur, Uttar Pradesh, India (2007)

This example (Figure T10) of a Gender access and control profile combines animal-related services and resources with management activities such as mixing feed and washing the animal. This profile was developed by a group of men who own working animals, as part of a situational analysis with the community (Chapter 4, Phase 2). The chart shows that women have access to most resources and services (10 out of 14) but only limited control over them. Men carry out several management activities, such as bathing animals, visiting the farrier, providing or seeking treatment and taking the animal out to graze. The group discussed the reasons for this and the effect of different access and control patterns on their animals' welfare.

T11 Changing trend analysis

What it is

Changing trend analysis is a matrix tool which helps the community to identify changing trends over time. Analysis of how change happened from past to present and what caused the changes enables you and the participants to understand their current situation more clearly.

Purpose

Changing trend analysis (Figure T11a) may be used to analyse many different aspects of peoples' lives and the lives of their working animals. Examples include how animal populations, feeding practices, disease patterns, availability of health services and availability of grazing land have changed over a period of time. The tool can show changes in the number of animals affected by disease or changes in the severity of the disease over time. This exercise can be good for starting to look at the causes of welfare problems in working animals and searching for solutions.

A variation is called Before-and-after analysis. This tool is frequently used with the community to analyse the care of working animals before and after implementation of their action plan to improve welfare. Village groups have analysed issues such as animal feeding, health service provision, harness repair, the number of times that animals are given water in a day, and many others. Before-and-after analysis is often used in the self evaluation phase of collective action (Chapter 4, Phase 6).

How you do it	
Step 1	Ask the group about the present situation relating to care of working animals in the community. Then ask how things have changed over time. Enable participants to list the things that have changed and to select the ones they feel are most important.
Step 2	Ask people to decide on a timescale for their analysis. This may be before and after you worked with the group to facilitate an animal welfare intervention. Or it may be based on other important activities, events or years. Some of our groups analyse animal management trends over generations, such as 'In our grandfathers' time, in our fathers' time and in our own time' (see Figure T11b).
Step 3	Ask participants to draw a matrix on the ground and show the time scale on the horizontal line. Depict the items for analysis on the vertical lines using symbols, seeds or pictures. With a Before-and-after analysis it may be easier to do this the other way around, showing time on the vertical axis. Decide with the group how they will display the situation, for example by scoring using seeds or stones, or by using symbols, leaves or sticks of different sizes.
Step 4	When the matrix is filled and the participants are satisfied, discuss the findings and trends over time. Ask why the changes happened and whether they are positive or negative changes. Explore possible options for action to continue with positive changes or reduce negative changes, carried out either by the group themselves or with support from outsiders.

Facilitator's notes: Changing trend analysis

- Always encourage people to analyse changing trends in depth. Allow enough time for all participants to remember and discuss their experiences
- Clarify any doubts about abrupt changes and understand people's perceptions about changes over the period. Discuss how things can be changed according to the community's aspirations.

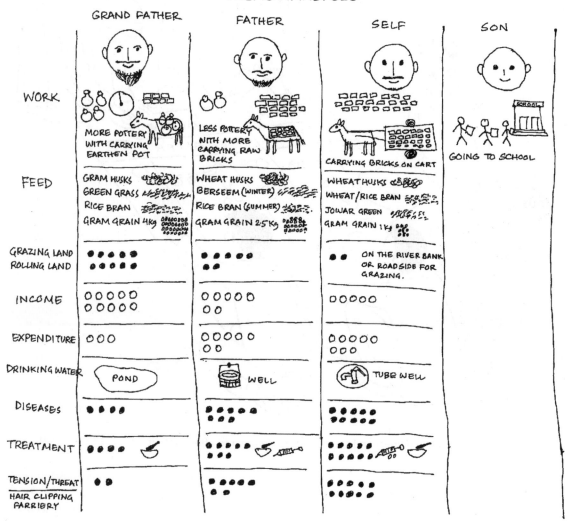

Figure T11a Changing trend analysis of changes affecting potters and their working animals over four generations in India (2008)

This matrix (Figure T11a) is the result of a Changing trend analysis carried out with a group of potters in Basantpur Sainthali village, Uttar Pradesh, India. It was used as part of a situational analysis (Chapter 4, Phase 2) and shows changes in work load, quantity and type of animal feed and fodder, grazing land availability, income and expenditure, animal disease prevalence and treatment options. The main profession of pottery has changed over the years and the group has become more dependent on their animals working in brick kilns. In the early days they were dependent on traditional treatment (pestle and mortar symbol) but gradually they are using modern veterinary treatments (syringe symbol) as well.

Figure T11b A group of animal owners and carers doing a Before-and-after analysis

T12 Dependency analysis

What it is

Dependency analysis helps to analyse a community's dependency on particular resources and service providers to sustain their livelihoods. It is adapted for animal welfare from an earlier version (Jayakaran, 2007).

Purpose

In the context of animal welfare, this tool can enable owners to analyse their control over the resources and services necessary to fulfil the basic needs of their working animals, such as dealing with a specific disease or providing water and good quality feed. It enables people to look at which provisions are within their own control (sometimes called internal control), which are controlled by other stakeholders or service providers (outsider control) and which are beyond the control of either owners or local stakeholders.

The first stage of Dependency analysis (Figure T12) is to identify the resources and services needed by working animals. This provides an opportunity to explore the group's perceptions and beliefs about the animal management and work practices which will prevent welfare problems from occurring in the first place. Analysis of their dependency on others encourages the animal owners to plan strategies for accessing resources and services which are not under their direct control, in order to improve their animals' welfare.

How you do it	
Step 1	Ask participants about the problems they face in managing their animals during work and rest. Alternatively you can start by asking about the needs of their animal. Draw a large circle on the ground, with a picture of the animal at the centre and show all these issues on the outermost periphery of the circle, using pictures or symbols.
	Then draw two more concentric circles inside the initial circle and divide the circles into segments, one for each problem or need shown
Step 2	Ask the group to indicate which of the animals' needs or problems that they can deal with themselves, which needs are met by outsiders (service providers) and which are beyond the control of both owners and outsiders.
	Then ask participants to score each need according to their own influence over it. For the first need or item ask: How much influence do you have over providing the need? How many seeds or stones (out of ten) would show your level of influence? On the outside circle, ask them to place the number of stones in each segment which reflects their own control over that particular problem or need.
Step 3	Discuss which needs are met with the help of outsiders (service providers). When do the group need to depend on outsiders to deal with their animals' problems or needs? Examples might include farriers (for shoeing), feed sellers or local animal health service providers. Out of the remaining seeds or stones, ask the group how many would represent the influence of outsiders (service providers) over provision of that need or item. Show this by placing stones in another circle inside the first one (see Figure T12 below).
Step 4	Finally, the remaining seeds or stones are placed in the innermost circle, indicating to what extent the provision of each animal need is beyond the control of either the animal owners or local service providers.
Step 5	When scoring is complete, begin a discussion with the group. Encourage them to search for ways to gain more influence or control over meeting their animals' needs or solving animal welfare problems themselves. Ask: How can the scores in the outside circle be increased? For each problem or need, analyse the factors responsible for dependency on others and the scope for reducing dependency on outsiders and increasing their own control. Include the items that are beyond anyone's control, such as diseases, disasters or environmental issues. Enable the group to discuss options for reducing the effect of these items on the welfare of their animals. Agree to take action, either individually or collectively.

Facilitator's notes: Dependency analysis

- This exercise can take considerable time, so discuss this in advance with the group and agree on a suitable time to set aside for doing it.
- The same exercise can also be shown as a scoring matrix rather than a segmented circle, using vertical and horizontal lines to divide the needs or problems and the levels of dependency.

DEPENDENCY ANALYSIS

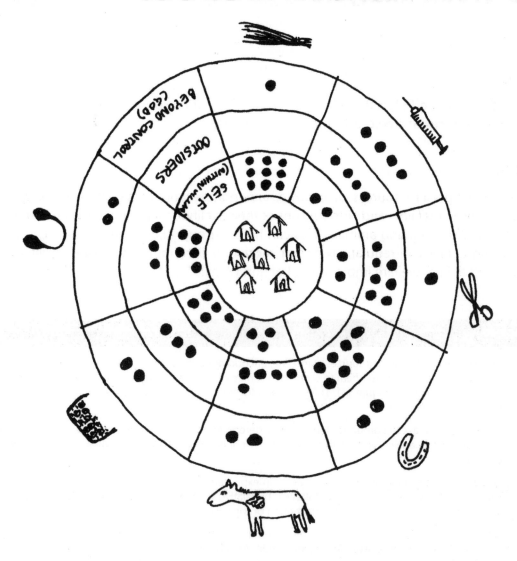

Figure T12 Dependency analysis by a community group at Bijouli village, Meerut, Uttar Pradesh, India (2008)

Figure T12 is the output from a Dependency analysis carried out by a community group in Bijouli village, Meerut in Uttar Pradesh, as part of a situational analysis (Chapter 4, Phase 2). The group listed seven needs of the animal on the ground using symbols: stable cleaning, veterinary treatment, hair clipping, farrier's services, early identification of disease, feed, and harness repair. The group felt that stable cleaning was almost entirely under their own control. They perceived that they were dependent on others for treatment and that sometimes even accessing treatment was beyond their control. This led to a lot of discussion on measures that could be taken to prevent disease and injuries. The group also found that they were more dependent on the farrier and hair clipper than on the harness-maker and feed seller, but still wanted to reduce their dependency on the feed seller, as food was a daily need for their animals. They decided on collective action to reduce this dependency by bulk-buying and storing feed as a group.

T13 Credit analysis

What it is

Credit analysis (Figure T13) is the collective analysis of existing sources of income, expenditure and different sources of credit, using scoring methods.

Purpose

The analysis of present sources of income, expenditure and credit enables the group to understand the constraints and compulsions of their livelihood situation and how these affect the welfare of their working animals. This can help a group to explore alternative sources of credit through collective regular savings or monetary contributions towards their self-help initiatives and animal welfare improvement action plans.

How you do it	
Step 1	Ask participants about their sources of income, encouraging them to depict all the types and sources of income they have. Then ask them to score each source of income according to its importance, using 10 stones or pebbles (see Figure 13). This generates further discussion and a better understanding of the income situation.
Step 2	In a similar way, ask the group to depict their different items of household expenditure. After all items have been listed, again ask them to score each item of expenditure relative to their biggest household expense.
Step 3	Discuss what they do when income is less than expenditure, or during a period of crisis. Ask them to show their various sources of credit, followed by relative scoring of each source of credit based on the source from which they borrow the most often or the largest amount.
Step 4	After the diagram is complete, initiate a more in-depth discussion on the advantages and disadvantages of each source of income, expenditure and credit. Ask about issues such as interest rates, availability of credit and ease of access to credit. Based on these discussions the group can search for alternative credit options and this may motivate them to start their own savings group to help to pay for animal welfare improvements.

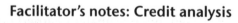

Facilitator's notes: Credit analysis

- For some groups this might be a sensitive discussion. Trust is important between group members before they will have an open discussion on income, expenditure and credit.

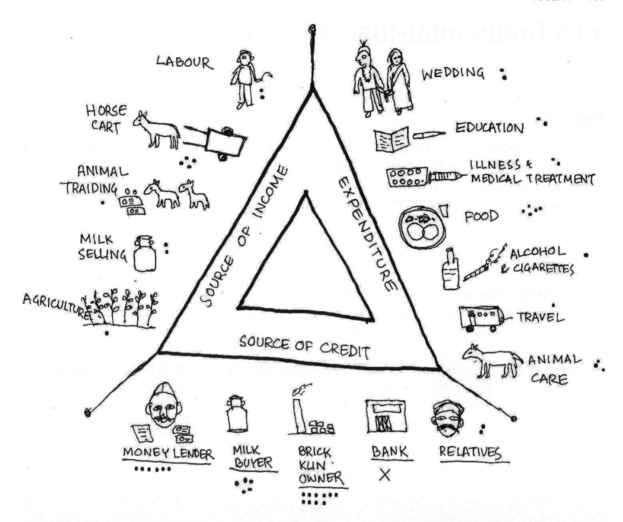

Figure T13 Analysis of income, expenditure and credit by horse owners in Subtu village, Muzaffarnagar district, Uttar Pradesh, India (2008)

Figure T13 shows the output of a Credit analysis exercise by a group of animal owners in Subtu village, Muzaffarnagar district in Uttar Pradesh, India, carried out as part of their situational analysis (Phase 2, Chapter 4). The diagram shows that they earn their main income though work with their horse cart. Additional income comes from selling buffalo milk and working as agricultural labourers. A few members earn money from animal trading and agriculture as well. Their major expenditure is on food for family members, and animal feed and care. Ninety per cent of the animal owners are dependant on their employers at the brick kilns for credit, and through loans from local moneylenders. Discussion during this exercise led the group to start their own savings fund in order to support their animal-related and household needs and reduce their dependency on loans.

T14 Group inter-loaning analysis

What it is

This tool provides a visual representation of the reasons for lending or borrowing money between members of the group. The group's savings fund or common contributions are lent (inter-loaned) for various purposes, such as buying animal feed, paying for services from different service providers, cart repair and maintenance, purchase of a working animal, and other household needs.

Purpose

Group inter-loaning analysis (Figure T14) provides you and the group with insight into the reasons for lending or borrowing money between its members. This can help participants to decide where collective spending might save them money and therefore reduce their need to borrow from the fund. For example, if many group members are borrowing money to buy animal feed, a common fund could be used to buy animal feed in bulk. The group may wish to arrange vaccinations for all village animals at the same time for a reduced fee. The same exercise can be used to assess the present status of repayment of loans and create peer pressure to repay the loans to the group fund as agreed between the group members.

How you do it	
Step 1	For this exercise the animal owners' group must have been meeting for at least one year, with members making regular contributions to a common savings fund. They must also have started lending money from the fund to the group members (inter-loaning).
	Explain the exercise and start with a general discussion, such as asking questions about the group and its activities.
Step 2	Encourage participants to discuss the collective fund they have generated and the amount saved by each member of the group. Ask them to make a big circle on the ground or on chart paper and mark the central point of the circle. Enable them to write the names of the group members on separate cards, or to identify group members using different symbols. Arrange the cards representing each person who has taken a loan around the outside edge of the circle.
Step 3	Draw three inner circles and divide the circle into as many parts (segments) as the number of group members who have borrowed from the common fund. In the second circle indicate the purpose for which the loan was taken and in the third circle the amount of money borrowed (see Figure T14). All of these can be depicted using symbols and stones, or may be written on cards. If any member has taken a loan twice this should be reflected in their column or segment.
Step 4	If participants wish, use additional circles to show how much of the loan has been repaid to the common fund and the amount of interest paid. Enable them to depict the number of instalments already repaid. Discuss the number of instalments remaining and the number of months until the final settlement will be made.
Step 5	Encourage participants to discuss their need for further loans, the purposes for which they will borrow money and any action that they can take to reduce their need to borrow money. At the end of the exercise, ask the group to summarise their findings and highlight where loans have been used for animal welfare improvement.

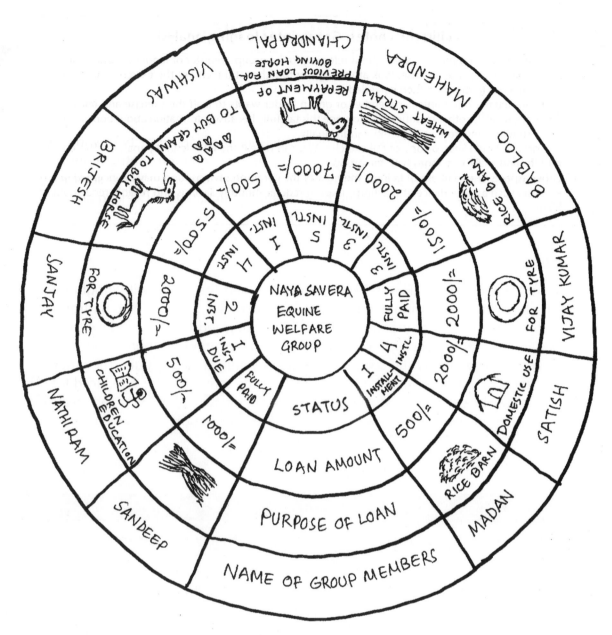

Figure T14 Group inter-loaning analysis carried out by Naya Sabera Equine Welfare Group, Bhakla village, Saharanpur, Uttar Pradesh, India (2008)

As part of their situational analysis (Phase 2, Chapter 4), members of the Naya Sabera Equine Welfare Group in Bhakla village, Saharanpur, Uttar Pradesh, India, analysed the use of money lent from their common savings fund. All members of the group had taken a loan at some point. The central circle in Figure T14 represents the status of the loan and number of installments still to repay. The second circle shows the amount of each loan. The third circle shows the reason for the loan; most loans are taken to buy working horses, buy animal feed or make repairs to the cart. This exercise helped the group to discuss where they could take action collectively for their animals, such as buying feed in bulk at a discounted price.

Facilitator's notes: Group inter-loaning analysis

- It is best to do this exercise without referring to the group's register or ledger of savings and loans. If any queries arise, look at the register together at the end of the exercise and make changes to the chart then if needed.
- Sometimes one particular member or group leader wants to lead the exercise and tell you about all the other members. Discourage this tactfully because information about borrowing should come from each member individually.
- The group should keep a copy of the exercise. You can repeat the inter-loaning analysis after 6 to 12 months and analyse changes with the group.
- Other aspects or conditions of being a member of the common savings fund may be included in this exercise, such as attendance at meetings, making regular monthly savings and reliable repayment of loans.

T15 Cost-benefit analysis

Animal welfare practices, animal-related service providers and animal feeding practices

What it is

This is a method of comparing the animal welfare and financial costs and benefits of a provision or resource.

Purpose

This tool explores the potential benefits, risks and affordability of various animal welfare-related activities. It can show costs and benefits to the animal and also to its owners, users and carers. The exercise motivates the group to take action to improve the welfare of their working animals and we use it in many situations, such as to look at the costs and benefits of preventive animal health measures (such as vaccination), the use of different veterinary services and the most cost-effective combination of animal feeds.

In our experience, animal-owning groups will usually start by analysing financial costs and benefits, looking for the cheapest options. Your role as a facilitator is to ensure that animal welfare costs and benefits are included in the discussions, so that reduction in expenditure does not lead to reduction in animal welfare. The group may monitor whether their changes improve or reduce animal welfare using the animal-based welfare indicators developed in other exercises, such as Animal body mapping (T20), 'If I were a horse' (T17) or How to increase the value of my animal (T18).

Cost-benefit analysis of animal welfare practices

This version helps the group to decide whether to use a particular resource or service, such as grazing areas, water sources, vaccination, farriery or hair clipping. For example, a cost-benefit analysis on veterinary treatment may analyse the cost of paying the vet and buying medicines against the welfare benefits to the animal and the family's loss of earnings if their animal is not treated. This is described in Case study M on tetanus vaccination.

Cost-benefit analysis of animal-related service providers

This looks at whether the use of one particular service-provider has greater costs or benefits than another. It may be done by comparing different services against each other, or by comparing people who provide the same service. (See Figure T15a)

Feeding practice analysis

This special adaptation of the Cost-benefit analysis tool explores whether changes in animal feeding practices will lead to a better outcome for animal welfare while remaining affordable. It can assess current feeding practices, the nutritional content of different feeds and their effect on working animals, the availability of alternative feed sources and the best formulation for balanced animal feed at low cost. Animal feeding practice analysis combines more than one tool and is described in detail in T27.

How you do it

Cost-benefit analysis is relatively simple when looking at resources that can be supplied by the community themselves. When analysing costs and benefits of animal-related service providers, aspects such as their location, distances from village households or quality of service make the analysis a bit more complicated.

Step 1	If the group would like to assess the costs and benefits of using different service providers, start by asking who provides all the services they need to look after their working animals. Draw a matrix and list the service providers along the top of the matrix. Examples could include the veterinary doctor, community animal health worker, foot trimmer, animal feed seller, harness- or cart-maker, medical store or agrovet supplier (see Figure T15a below).
Step 2	Discuss each service and ask the group to list factors that could be seen as costs or benefits of the service provider. These could include: their distance from the village, costs involved, mode of payment, quality of service and any other relevant issues, such as how well the service provider handles the animal. List these factors down the side of the matrix.
	Alternatively this can be done the other way around, with factors listed along the top of the matrix and service providers listed down the side.
Step 3	Facilitate the group to fill in the matrix for each service provider, as shown in Figure T15a.
Step 4	Once completed, ask the group to analyse the costs and benefits of different service providers. What are the possible alternatives when costs are high? Can the benefits be increased in order to improve animal welfare? Can the costs be reduced without a negative effect on animal welfare? Based on these questions the group can decide on what action they would like to take.

Facilitator's notes: Cost-benefit analysis

- If you are discussing animal treatment services, ask the group about the occurrence of diseases during the last one or two years, the treatment services they used, duration of recovery period, costs involved and losses incurred when the animal could not work.
- The presence of technical people such as vets or community animal health workers can de-mystify diseases and their treatment, enabling the community to discuss them in more depth. This also helps the technical person to understand the community's perspective, bringing together local wisdom with experts' views and experiences.

	LOCAL HEALTH PROVIDER	FERRIER	HAIR CLIPPER	FEED SELLER	CART REPAIRING	MEDICAL STORE	SADDLER
WHERE	Fauji Shyamli	Dhole Stand Shahpur	Shahpur	Sisoli	Savatu	Sisoli	Sisoli
DISTANCE	25 km	7 Km	12 Km	4 Km	0 Km	4 Km	4 Km
COST (IN 3 MONTH)	400/=	720/=	180/=	6390/=	70/=	50/=	300/=
MODE OF PAYMENT	Cash	Cash	Cash	Cash	Cash	Cash	Cash
EFFECT ON ANIMAL	NO-INCOME ANIMAL WILL BE OF NO USE	Lameness	Improper Sweating leading to Illness	Weakness leading to Disease	More Power needed to Pull Cart, suffers	No Timely Treatment	wounds
QUALITY	Good	OK/Good	Very Good	OK	Good	OK	Good
IMPORTANCE (OUT OF 5)	●●	●●●●	●	●●●●	●	●	●

Figure T15a Cost-benefit analysis of animal-related service providers in Subtu village, Muzaffarnagar district, Uttar Pradesh, India (2008)

The animal owners of Subtu village, Muzaffarnagar district, Uttar Pradesh, India, analysed the costs of animal-related service providers in terms of distance from the village, monetary costs and costs to their animal (animal welfare effects) if the service was not used (Figure T15a). The last two rows depict the quality of the service provider and his or her importance for their working animals. After drawing the diagram the group discussed their options for reducing costs and improving the quality of these services through collective action. They worked out how to engage each service at a reduced rate for the group and decided to maintain an animal first aid kit in the village to reduce expenses at the medical store (Figure T15b below).

SERVICE PROVIDER	PRESENT MONTHLY COST	WAY OUT OF REDUCING COST
L.H.P COMUNITY HEALTH SERVICE PROVIDER	400	• WE WILL FORM A ANIMAL SERVICE CENTRE AND CALL THE COMMUNITY HEALTH SERVICE PROVIDER COLLECTIVELY
FERRIER	720	• WE WILL COLLECTIVELY CALL THE FERRIER TWICE, IN THE VILLAGE / TONGA STAND AND MONTHLY FIX RATE THE RATE FOR FERRIERING.
		• WE WILL MAKE AN AGREEMENT REGARDING USED/OLD SHOES
		• WE WILL CALL ONLY ONE FERRIER
HAIR CLIPPER	180	• LIKE FERRIERING, HAIR CLIPPING WILL ALSO BE DONE COLLECTIVELY TO REDUCE THE COST.
FEED SELLER	6390	• COLLECTIVE PURCHASING OF RICE BARN, WHEAT STRAW ETC. COLLECTIVE PURCHASING OF • MAIZE, WHEAT & BARLEY AND ITS GRINDING.
		• WE WILL NEGOCIATE AND FIX THE PRICE OF WHEAT STRAW BEFORE HARVESTING.
CART REPAIRER	70	• DAILY USE OF GREEN FODDER
		COLLECTIVELY NEGOTIATE FOR BETTER PRICE
MEDICAL STORE	50 —	WE WILL START MAINTAINING FIRST AID KIT WITH US.
SADDLER	300 —	• COLLECTIVELY NEGOTIATE FOR BETTER PRICE
		• PROPER SETTING OF SADDLE
		• WE WILL GREASE THE SADDLE ON WEEKLY BASIS.
		• TIMELY REPAIR OF SADDLE

Figure T15b Village action plan to improve animal-related services in Subtu village, Muzaffarnagar district, Uttar Pradesh, India (2008)

Case study M. Cost-benefit analysis of tetanus vaccination in Mhallaur Anshik village

Source: Vineet Singh and Ram Raksh Pal Singh, Arthik Evam Jan Kalyan Sansthan, Lucknow, Uttar Pradesh, India, June 2008

A community in Mhallaur Anshik village carried out a Cost-benefit analysis to find out if it was beneficial to vaccinate their working horses against tetanus. After concluding that the benefits outweighed the costs, they explored whether it was possible to reduce the costs of vaccination. The group consulted their local animal health provider to gain a better understanding of what was needed. They agreed to vaccinate their animals as a group and to buy the vaccines in bulk from the nearest city. Money was collected in advance, including travel expenses to the city. They negotiated a discounted price with the animal health provider if he vaccinated all the horses, including booster vaccination. The community calculated that by doing this they had saved a considerable amount of money and helped all of their animals to avoid suffering from tetanus. The illustration below shows their financial cost-benefit calculations.

COST BENEFIT ANALYSIS

KAMLESH'S COST (WHOSE MULE DIED OF TETANUS)	GHANSHYAM & TEKRAM'S COST (WHO GOT THEIR ANIMALS VACCINATED)
1) PURCHASE COST OF MULE - 14500/=	COST OF VACCINATION OF ONE ANIMAL (VACCINE, SYRINGE & VET'S FEES) — 16/=
2) COST OF TREATMENT (MEDICINE & VET'S FEES) - 700/=	
3) LOSS DUE TO NO WORK FOR 3 DAYS (500/= × 3) - 1500/=	TOTAL COST OF VACCINATION OF 32 ANIMALS 16 × 32 = 512/=
TOTAL LOSS - 16700/=	TOTAL COST OF VACCINATION 512/=

T16 Animal welfare snakes and ladders game

What it is

Our community facilitators have found the Cost-benefit analysis tool (T15) to be an effective way to motivate groups to improve the welfare of their working animals. They came up with different ways to do it in a more fun and entertaining way: one of these is an adaptation of the traditional 'snakes and ladders' game.

Purpose

The Animal welfare snakes and ladders game engages the interest of participants, increases their collective knowledge about animal management and motivates them to act on poor welfare practices. Both adults and children enjoy this game.

Figure T16a Traditional snakes and ladders game large enough for players to walk around

How you do it	
Step 1	For this game you need to prepare beforehand.
	The common Snakes and Ladders board game is used for the exercise, either the small size available in the market, or you can make a big version using large sheets of cloth or paper. The game has between 50 and 100 squares in a matrix. Ladders and snakes are drawn or painted on, connecting different squares (see Figure T16a and b)
	To convert the game for animal welfare cost-benefit analysis, two types of information need to be collected before starting: • Existing animal management or work practices which are positive/good • Existing animal management or work practices which are negative/bad
	These practices need to be recorded in advance on cards, using words, symbols or photos. Place one card in the square at the top and bottom of each snake, and one at the top and bottom of each ladder. Snakes are usually associated with cost or loss and ladders with benefit or gain. This cost or loss and benefit or gain can be expressed in both welfare and financial terms. For example, bad hoof care practices can lead to lameness (pain and poor welfare for the animal) as well as three days' loss of work for the owner, costing 300 Indian rupees or 30 Egyptian Pounds a day. Alternatively, causes and effects can be put at opposite ends of the snakes and ladders. For example, the card at a snake's head could show a lame animal, while the one at the tail could show poor hoof care practices (bad quality shoeing or untrimmed feet).
Step 2	Normally six to eight people play the game together using dice. Each participant is represented by a different counter or object (such as a bottle top, leaf or stone) placed on the first square. Everyone gets a chance to play by rolling the dice in turn. At the beginning somebody has to roll a six before the game can start. Then each player moves their counter the same number of squares as he or she rolls on the dice.
Step 3	When a player's counter lands on a square containing a card or photo, the card is turned and discussed by the players. If a player reaches a square showing the head of a snake, the counter must be moved down to the tail of the snake. When a player reaches a square at the bottom of a ladder they can climb the ladder to reach the square at the top.
	Before moving from a snake's tail, players must describe a situation that they have experienced which is similar to the one shown on the card. Encourage the group to discuss this and decide what types of action would turn the cost into a benefit, before moving on to the next player.

Facilitator's notes: Animal welfare snakes and ladders game

- We have used many variations of this game, such as placing question cards about good and bad animal management practices in random squares on the board.
- The board can be made so large that players can walk around it to play.
- We have often used this game with children, with great success.

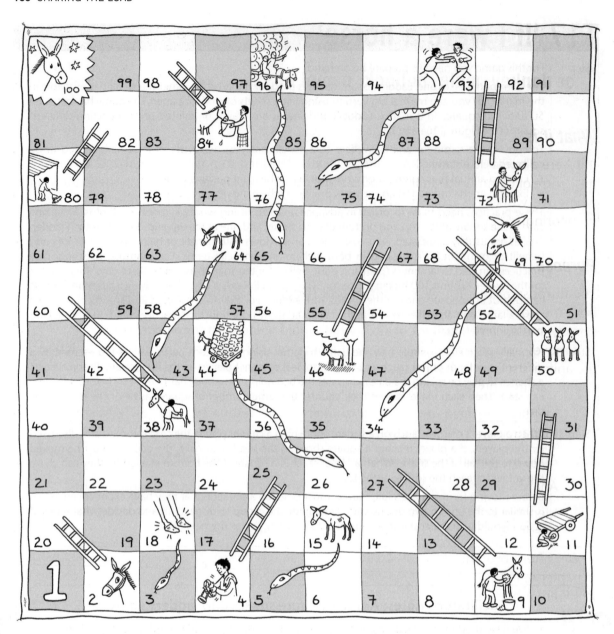

Figure T16b Snakes and ladders game for animal welfare sensitization, illustrated example from an original large cloth board, developed by Brooke Egypt, Cairo (2009)

T17 'If I were a horse'

... or donkey, mule, bullock, camel or yak

What it is

'If I were a horse' is a new PRA tool specifically designed to put the animal and its welfare at the centre of community analysis and discussion. This tool is very popular in the communities where we work. It is used to identify animal welfare issues for intervention planning and monitoring.

Purpose

The purpose of the 'If I were a horse' tool (Figure T17) is to enable people to visualise life from the point of view of their working animals, asking the question: 'If you were a horse (or donkey, mule, camel, bullock or yak), what would you expect from your owner?'

It moves owners from looking only at animal-related resources and services, to looking at the animal itself and what its appearance and behaviour tells them about its welfare.

You can use this tool to help the group to:

1. Identify the needs of working animals – the resources, services, environment and management practices that will enable them to have good welfare.

2. Analyse how far those needs are met by their owners and other service-providers

3. Analyse the effects on working animals when their basic needs are not fulfilled

4. Identify signs visible on the animal or behaviour that the animal shows when each of its needs are being met or are not being met (animal-based welfare indicators).

We often use this tool as the first step in participatory welfare needs assessment (Chapter 4, Phase 3). The animal-based welfare indicators identified by the group during the If 'I Were a horse exercise' can then be used as a basis for the Animal welfare transect walk (T22).

How you do it	
Step 1	This exercise starts with a question to the group: 'If you were a horse (or donkey, mule, camel, bullock or yak), what would you expect from your owner?' This encourages them to try to see the world from an animal's point of view.

Draw a big circle on the ground or paper with a working animal in the centre. If you carry drawings or models of animals with you, such as the horse jigsaw puzzle used for Animal body mapping (T20), one of these can be put in the centre of the circle instead.

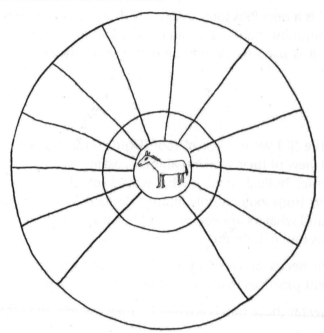

Show all the expectations that the group thinks animals have of their owners, using words, pictures or symbols on the inner periphery of the circle. In the illustration below, owners chose to write expectations in the inner circle and drew a corresponding symbol on the outer periphery of the circle.

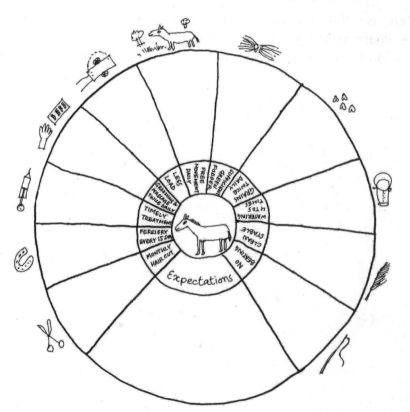

Step 2	Ask participants how far they think their animal's expectations are currently fulfilled. Draw another circle outside the first one and ask the group to score the extent to which their animals' expectations are met, using up to ten stones, beans or seeds. 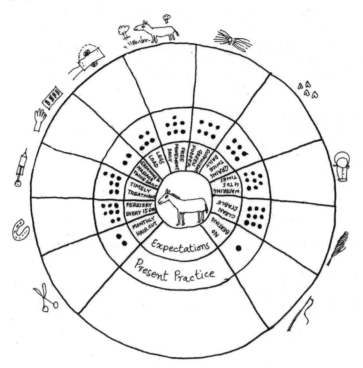 Here we find that people usually score according to the how much they can currently afford or make available to their animal. With the group, discuss and analyse the reasons for the low scores (animals' expectations which are not well met).
Step 3	Once all the expectations are scored, start a discussion about the effects on the animal when its expectations are not fulfilled (when the issue scores less than 10). Ask the group: 'If the expectation is not fulfilled what effect would this have on the animal?' Draw another circle outside the first two. For each expectation, show these effects using symbols or by writing on cards and placing them on the circle. 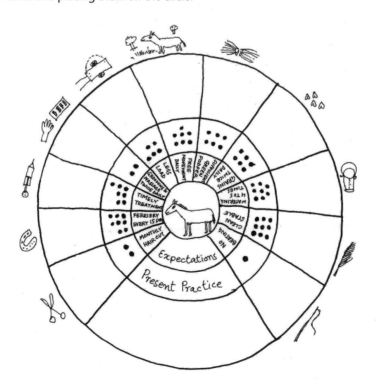

Step 4	Make a fourth circle outside the others. This is used to identify how each effect from Step 3 would be seen on the animal, either as physical signs on its body or in the way it behaves (see Figure T17). Facilitate the group to analyse these behavioural and physical signs very thoroughly. What are the signs on the animal if its expectations are not met at all, or if they are met to a small extent? How do the signs change if the expectations are met to a greater extent or if the needs are fully met?
Step 5	Ask the group how all these behavioural and physical signs (animal-based welfare indicators) can be measured and recorded, including where to look and what to look for. Discuss the importance of measuring these signs and how this can help people to understand what their animals are feeling and what they expect or want from their owners.
Step 6	Ask the group to list all the behavioural and physical signs from Step 4 as symbols, pictures or words on chart paper, or in a ledger or register. Include the decisions on how they will be measured. Agree a time when the group will assess all their animals by doing an Animal welfare transect walk (T22) together.

Facilitator's notes: 'If I were a horse'

- Encourage everyone to express their own views and avoid using only one person's examples or contributions for the diagram. Allow enough time to discuss participants' own beliefs and traditional animal management practices.
- Take notes of important management and work practices for further discussion and action.
- As an alternative to the diagram shown above, you can begin by showing the animals' expectations using pictures or symbols on the outer periphery of the circle and continue the exercise toward the centre of the circle.

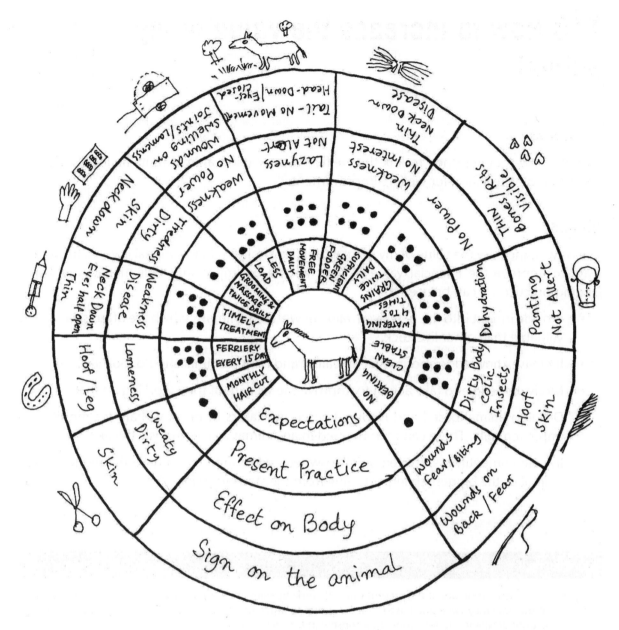

Figure T17 'If I were a horse' diagram from Burana village, Muzaffarnagar, Uttar Pradesh, India (2009)

Figure T17 shows the completed 'If I were a horse' diagram developed by a group of animal owners in Burana village, Muzaffarnagar, Uttar Pradesh as the first step in a participatory welfare needs assessment (Chapter 4, Phase 3). They identified eleven expectations that their animals have of them as owners: no beating, a clean stable, watering 4 to 5 times a day, grain twice a day, supplementary green fodder, free movement (loose grazing) daily, less load, grooming and massage twice a day, timely treatment, farrier every 15 days and a monthly hair cut (to keep them cool). In the second circle they scored their present practices out of ten: they scored lowest on not beating and highest on stable cleaning. Through discussion the group identified the effects of not meeting their animals' expectations, including fear, lameness, weakness and wounds. Finally they identified where they would look for the specific behavioural and physical signs resulting from not meeting the expectations of their animals. These were recorded and used by the group to assess their animals during an Animal welfare transect walk (T22).

T18 How to increase the value of my animal

What it is

This is a new tool designed to link the welfare of a working animal to its financial value, in order to motivate action for welfare improvement.

Purpose

The How to increase the value of my animal tool (Figure T18a) helps animal owners to identify ways to improve the financial value of their animal by improving their management and feeding practices. It can be used to:

- Provide insight into the perceived value of working animals, which are often seen as less valuable than other livestock.

- Increase knowledge about what a healthy and happy animal should look like.

- Motivate people to take action to improve the welfare of their working animals.

- Develop animal-based and resource-based indicators of good welfare for monitoring improvements. These can then be used in an Animal welfare transect walk (T22).

- Initiate competition between animal owners, who often try to have the highest value animal or the biggest change in value from the starting point.

How you do it	
Step 1	Start a discussion about the present value of working animals belonging to the group and their value at the time they were purchased. Ask the group to agree on values collectively. This usually involves a lot of debate in order to come to a consensus for each animal. Ask participants to list the name of each owner and the present value of each of his or her animals on a chart or a piece of paper (see Figure T18a, Step 1).
Step 2	Facilitate the group to identify which animals have the highest and lowest value and ask about the reasons for the values given. Participants may mention age, breed, size, character, health and other animal parameters, management or work-related issues and the preventive health or husbandry practices of the owner. List these on the chart or paper.
Step 3	When this list of reasons is complete, go with the group to visit each animal (this is an Animal welfare transect walk, T22). Encourage the group to agree the welfare status of each animal using the reasons or parameters that they have listed (see Figure T18a, Step 2). Discuss and agree the value of each animal again, based on the findings of the transect walk.
Step 4	Ask people individually about how much they would like to increase the value of their own animal in the future. Encourage each group member to decide on an action plan to increase the value of his or her animal, by improving the animal parameters and management practices identified. Finally the group should agree collectively on what the increased value of each animal would be if the action plan were implemented successfully (see Figure T18a, Step 3).
Step 5	Monitor progress on the individual action plans during each group meeting. Repeat the Animal welfare transect walk regularly with the group to see improvement in the welfare and value of each animal.

Figure T18a Steps in using the tool How to Increase the value of my animal

Case study N. Increasing the value of my working animal

Source: Dev Kandpal, Brooke India, Saharanpur, India, October 2008

In the village of Khurana, twenty people each own one horse which they use for transport of people and goods. After rapport-building by Siyanand, the community facilitator, the owners formed a Horse Welfare Group in 2007 and started to meet regularly. Siyanand wanted to motivate the group to improve the welfare of their animals and during one of their monthly meetings he initiated a discussion on the estimated value of their horses. This led to tremendous debate and discussion before the group came to a consensus on each animal's value.

Siyanand went on to ask the reasons for differences in value and why the cheapest and most expensive animals were given those values. That analysis enabled group members to describe which management practices increased the value of animals and which ones led to reduction in value. The group listed practices such as grooming, stable cleaning, hoof cleaning, hoof trimming, feeding, and beating. Siyanand then asked the question 'What can you see on the animal which tells you whether the owner is carrying out good management practices or not?' The group made a list of signs, including hooves trimmed and not cracked, eyes clean, shiny skin, no wounds on the girth and wither, and many others. They agreed that they could use their list to assess the present value of their animals. Then they went to see each horse and assessed its real status compared with the value they had given initially while sitting together, changing the value if necessary. Animals were given a score of good, medium or poor for each of the agreed signs. Once the assessment was completed participants made a summary, adding up the number of good, medium and poor welfare signs on each horse.

The group found that Balveer's horse, which was valued the highest, had six signs scored as 'good', eight as 'medium' and only one as 'poor'. The horse with the second highest value belonged to Manoj with ten body signs scored 'good', 'medium' and two 'poor' (see Figure T18b). This led each owner to analyse the welfare scoring system critically and think about its relationship to the value of his animal.

NAME OF OWNER	Present value of animal (Indian Rupees)	GOOD	MEDIUM	POOR
Prem	17,000	9	2	4
Balveer	25,000	6	8	1
Ram Kumar	18,000	6	5	4
Subhash	24,000	7	6	2
Balesh	11,000	7	4	4
Manoj	22,000	10	3	2
Pawan	11,000	7	7	1
Gurdas	18,000	3	6	6
Kuldeep	24,000	5	9	1
Herphool	10,000	7	0	8
Megh Raj	20,000	8	4	3

Figure T18b Analysis of animal welfare status compared to the present value of the animal, Khurana village, Saharanpur, Uttar Pradesh, India (October 2008)

Based on this, everyone agreed that they would like to increase the value of their own horse and decided the value that they would like it to have. Siyanand facilitated the group to help each person make specific action points that would increase the value of their animal. The group leader recorded all decisions and actions on chart paper and in the following monthly meetings the group used this chart to monitor their action points.

After six months the exercise was repeated: each horse was assessed again using the same body signs and its value was decided by the group. Owners saw an improvement in the welfare status of all the animals and valued all of them higher than before (see Figure T18c).

NAME OF OWNER	VALUE OF ANIMAL (INDIAN RUPEES)	
	Present Value	After 6 months
Prem	17,000	20,000
Balveer	25,000	28,000
Ram kumar	18,000	22,000
Subhash	24,000	28,000
Balesh	11,000	15,000
Manoj	22,000	25,000
Pawan	11,000	15,000
Gurdas	18,000	21,000
Kuldeep	24,000	25,000
Herphool	10,000	12,000
Megh Raj	20,000	23,000

Figure T18c Increase in the value of working animals due to welfare improvement, Khurana village, Saharanpur, Uttar Pradesh, India (October 2008)

This tool was named How to increase the value of my animal and has since been used to motivate many animal-owning groups to improve their animals' welfare. People's sensitivity towards their working animals and knowledge about animal needs also grows through peer pressure and through increasing the value of an important asset.

Facilitator's notes: How to increase the value of my animal

- We like to extend this exercise by encouraging the group to list the cost of the management changes needed to increase each animal's value and the time that will be taken to achieve this.
- Owners may also decide to have their action plans monitored by other owners in the group (sometimes three or four members will monitor all the actions).
- In our experience, groups review these action plans frequently, usually during every monthly meeting. They also physically monitor changes in their animals at least once a month by doing an Animal welfare transect walk (T22) because increasing the value of their animal is highly motivating.

T19 Animal feelings analysis

What it is

Animal feelings analysis is a new tool which looks at the working animal's experience of its own life. It analyses how the animal feels, or its 'mental welfare' (see Chapter 2). Examples include how happy, sad, relaxed, tense, depressed or frightened the animal feels, and how these emotions can be seen in its behaviour. This tool can be used on its own, or incorporated into the Animal welfare transect walk (T22).

Purpose

The Animal feelings analysis (Figure T19) has been developed specifically to bring out the mental aspect of animal welfare, because existing PRA tools do not cover this. Other tools in this section, such as 'If I were a horse' (T17) and Practice gap analysis (T21), look at the physical aspects of welfare such as food, water, shelter, disease and injury, and their effects on the animal's behaviour.

We have found that the Animal feelings analysis tool is very effective for:

- Sensitizing animal owners, handlers and carers to the fact that animals have feelings.

How you do it	
Step 1	The Animal feelings analysis is best carried out in small or medium-sized groups. Start by discussing what activities or types of work their animals do for them. Where and when do they use their animals for these activities? How much do they earn from these types of work?
Step 2	Ask participants what they do for their animals to keep them healthy and happy. Agree on the various animal expressions, body postures and behaviours that they look at in order to understand how their animals feel. An example given by our communities is that working mules use the position and movement of their ears, the movement of their eyes and the position of their head and neck to show if they are happy or sad.
	Encourage participants to describe as many different behavioural signs as they can.
Step 3	Categorize these behaviours into a matrix, showing the combination of behaviours that indicate a happy animal, those that indicate a sad animal, and those that indicate neither happy nor sad (neutral feelings). In our example (see Figure T19), a happy mule holds its head and neck above the level of its body, with ears forward and eyes moving. If the head and ears are in a medium position it has neutral feelings. If its head and neck are held below the level of its back, with eyes nearly closed and ears back, the mule is sad.
Step 4	This exercise can be done during a group meeting as a sensitisation exercise on animal feelings, or it may be incorporated as part of the Animal welfare transect walk (T22). If you are using it as part of the transect walk, first carry out steps one to three above. Then develop a matrix with the agreed behavioural signs written along the top and the names of owners and their animals written down the side. You can use either numerical scoring or traffic lights to indicate happy (green), medium/neutral (orange) and sad (red). Go with the group on a transect walk to visit individual animals and assess their behaviour. Discuss the body language of each animal in detail, to build consensus among all the participants about how each animal feels.
Step 5	At the end of the transect walk, sit together again and discuss which factors cause or influence the feelings of animals, both positively and negatively, and the reasons for this. Summarize the findings and decide on action points for individuals and the group in order to make their animals feel happier. Agree on a date to repeat the exercise and monitor changes in animal feelings.

- Demonstrating that animals express their feelings through their behaviour, or 'body language', in a way that people can understand.

- Identifying the signs and symptoms that people use to assess the condition of their animals.

- Analyzing positive and negative factors influencing the feelings and behaviour of working animals and discussing how to improve negative situations.

- Creating individual and group motivation to improve welfare.

Facilitator's notes: Animal feelings analysis

- Behavioural expressions or body language may be different for different species of working animal. There may be more than one type of behavioural expression for the same feeling. Encourage participants to discuss this.
- This exercise helps the group to appreciate that animals, like people, are sensitive and have feelings about what is happening to them at any particular time. It shows people that they can recognize how their animals are feeling. It motivates the group to care more about the effect of their own actions on their animals' feelings.

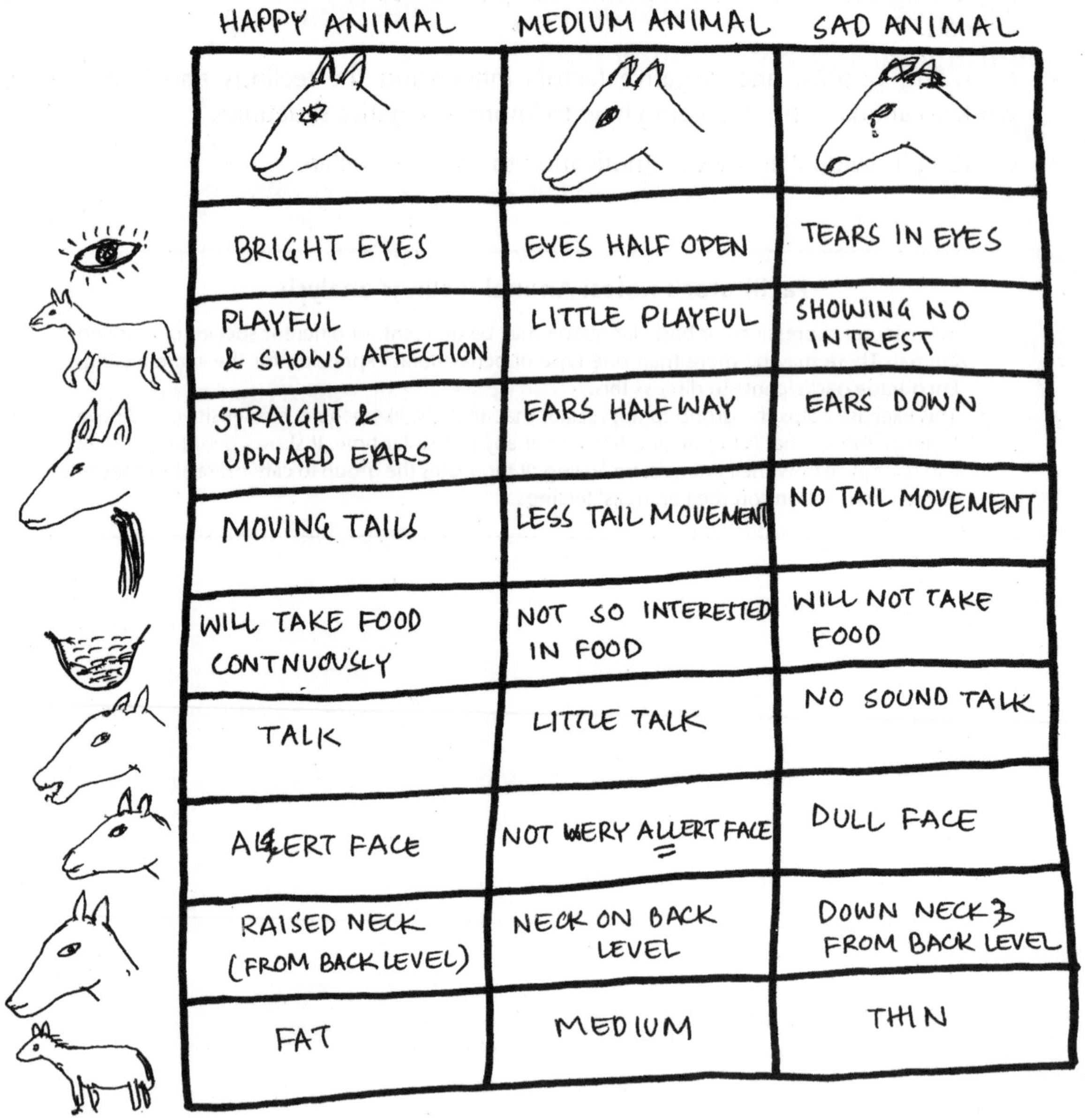

Animal Feeling Analysis

	HAPPY ANIMAL	MEDIUM ANIMAL	SAD ANIMAL
	BRIGHT EYES	EYES HALF OPEN	TEARS IN EYES
	PLAYFUL & SHOWS AFFECTION	LITTLE PLAYFUL	SHOWING NO INTREST
	STRAIGHT & UPWARD EARS	EARS HALF WAY	EARS DOWN
	MOVING TAILS	LESS TAIL MOVEMENT	NO TAIL MOVEMENT
	WILL TAKE FOOD CONTNUOUSLY	NOT SO INTERESTED IN FOOD	WILL NOT TAKE FOOD
	TALK	LITTLE TALK	NO SOUND TALK
	ALERT FACE	NOT WERY ALLERT FACE	DULL FACE
	RAISED NECK (FROM BACK LEVEL)	NECK ON BACK LEVEL	DOWN NECK 3 FROM BACK LEVEL
	FAT	MEDIUM	THIN

Figure T19 Animal feelings analysis by mule owners in Khurampur village, Ghaziabad, Uttar Pradesh, India (2008)

As the first step in Participatory welfare needs assessment (Chapter 4, Phase 3) a group of mule owners in Khurampur village, Ghaziabad, Uttar Pradesh, identified nine indicators that they could use to assess happiness and unhappiness of their animals. All animals were scored collectively based on these indicators, with three pebbles for 'happy', two pebbles for 'medium' and one pebble for 'sad'. Only one animal scored three pebbles for all the behavioural signs assessed. The animal owners discussed why this mule was found to be so happy and most of the other animals were not. Based on the discussion the group agreed on several action points to make the other animals happier.

T20 Animal body mapping

Animal body parts map, animal body wounds map, animal-based indicators body map

What it is

A body map is a picture of the body of a working animal showing the parts of the body, their functions, and the body areas affected by wounds or diseases. It depicts the animal's body or a particular part of the body as it is perceived by individuals or the community group.

Purpose

Body mapping is a useful exercise to explore the different perceptions that people have about their animals' body parts and the roles and functions of each part, because these perceptions will affect how they deal with wounds, diseases and other welfare problems. It can also be used to identify local names for the parts of the body.

We have included three different examples that we often use with the community:

Animal body parts map (Figure T20a)

This basic map looks at parts of the body and how they are perceived. It can be used to ask questions about how the group feels that a normal, healthy animal should look. We find this to be useful in places where all working animals have the same welfare problems, such as being thin, because owners may perceive this to be normal rather than unhealthy. The body map can initiate a discussion about what appearance is normal for a healthy animal.

Animal body wounds map (Figure T20b)

This map specifically indicates wounds or injuries and their causes.

Animal-based indicators body map (Figure T20c)

This exercise is often used as a starting point for discussion about prevention of welfare problems, and discussion of first aid measures for particular diseases or injuries. It can also be used to lead into other exercises, such as:

- Identifying how animal owners would assess problems on each body part (animal-based indicators of welfare), in preparation for an Animal welfare transect walk (T22)

- Looking in more detail at some of the problems identified, such as wounds and diarrhoea, using the Problem horse exercise (T25)

How you do it	
Step 1	Ask participants to draw an outline of the body of their animal on the ground or on paper. Identify the different body parts and the local names used for each body part. Initiate a discussion on the roles and functions of each part. This is an Animal body parts map
Step 2	Agree what to show on the body map depending on the focus of the discussion (see the variations above). For example, ask where people normally find wounds on the body, or how symptoms of particular diseases are seen on the body. Encourage participants to draw these on the body map or represent them using symbols next to the appropriate body part.

Facilitator's notes: Animal body mapping

- In our experience, it is best to use the Animal body mapping tool in a less crowded place so that onlookers do not distract the attention of animal owners.
- You can also use this tool to discuss traditional beliefs about the animal's body parts.
- In India we have adapted this exercise into a 'Broken Horse' jigsaw puzzle. We initiate discussion about the body of the animal using a wooden jigsaw of animal body parts, which owners put back together to make a complete picture of the animal. This puzzle is also very successful for starting discussions about animal welfare with children.

Figure T20a shows a simple animal body map drawn on the ground by ten animal owners from Jalalabad village, using sticks and coloured chalk powder. The discussion was initiated with the question: 'What parts of the body do you look at when purchasing an animal?' Once the body parts were identified by all the members of the group, they were encouraged to describe the health problems that could affect each body part. The owners showed all the places where wounds were found and indicated the causes of the wounds, producing Figure T20b – an animal body wounds map. In a later meeting the two maps were brought out again and the group discussed what they would see on each body part if the animal were in a poor welfare state. This exercise generated a long list of parameters for assessing animal welfare (Figure T20c) and was used to score all the animals during an Animal welfare transect walk (T22).

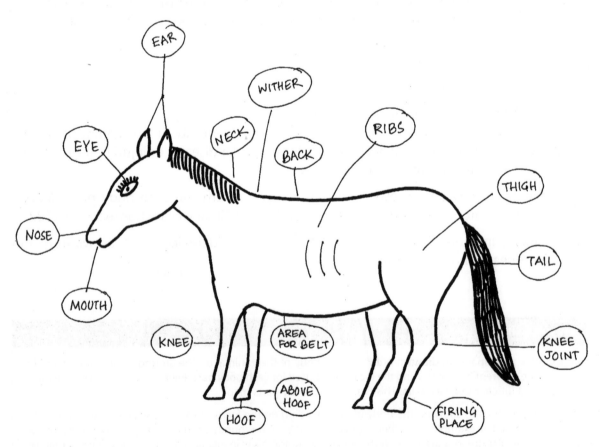

Figure T20a Simple animal body parts map of a working horse, Jalalabad village, Saharanpur, Uttar Pradesh, India (2007)

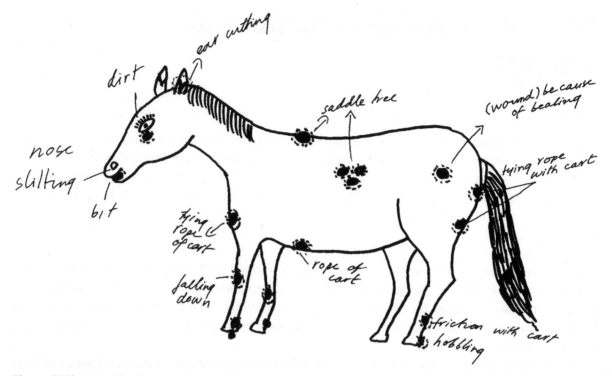

Figure T20b Animal body wounds map, indicating wounds and their causes on a working horse, Jalalabad village, Saharanpur, Uttar Pradesh, India (2007)

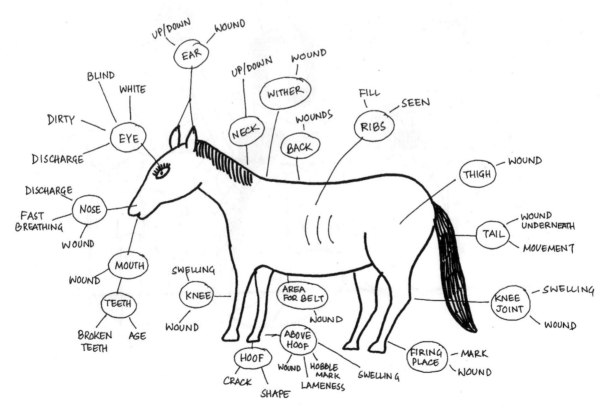

Figure T20c Animal-based indicators body map of a working horse, Jalalabad village, Saharanpur, Uttar Pradesh, India (2008)

T21 Animal welfare practice gap analysis

What it is

Animal welfare practice gap analysis is a tool designed to explore current animal management practices and activities which prevent working animals from experiencing poor welfare. It also identifies any gaps in these practices and reasons for the gaps. It has been adapted from other tools (Jayakaran, 2007) specifically to put the animal at the centre of the analysis.

Purpose

The first two steps are the same as in 'If I were a horse' (T17). However, in Animal welfare practice gap analysis the causes or reasons for not fulfilling the animal's expectations are analysed in much more depth. This helps the community to make more detailed actions plans for addressing the underlying causes of poor welfare by improving their animal husbandry and management practices.

Animal welfare practice gap analysis can be illustrated using a circle format (Figure T21a) or a matrix format (Figure T21b).

How you do it	
Step 1	Ask the group: 'If you were a horse (or donkey, mule, camel, bullock or yak), what would you expect from your owner?' This encourages participants to try to see the world from an animal's point of view.
	Draw a big circle on the ground or paper and draw the animal in the centre. If you carry drawings or models of animals with you, one of these can be put in the centre of the circle instead.
	Show all the expectations using pictures or symbols on the outer periphery of the circle (or on the innermost circle if you prefer to work outwards from the centre). Have a look at 'If I were a horse' (T17) for more detail.
Step 2	Ask the participants how far their animals' expectations are currently fulfilled by their current management and work practices. Draw another circle inside the first one and score how far the expectations are met using up to ten stones, beans or seeds.
	Here we find that people usually score according to the how much they can afford to make available to their animal. With the group, discuss and analyse the reasons for the low scores (the expectations which are not well met).
Step 3	After the scoring, initiate discussion and analysis of reasons for any gaps between the animal's expectations and the owner's actual management and work practices. This will bring out the reasons why the group cannot provide particular resources and services, or protect their animals from environmental or work factors which affect their welfare.
	Enable the group to identify the three or four most important factors (causes) responsible for the practice gaps. Put them into three or four more circles inside the first two (see Figure T21a).
	They may identify reasons such as: • 'We were not aware that animals had this expectation until we all started to discuss it'; • 'We don't know how to meet this expectation'; • 'We don't have the resources available to meet this expectation'. • 'We don't have time to meet this expectation'; • 'It isn't our habit to meet this expectation or we don't prioritise it', described in Figure 21a as 'carelessness'.
Step 4	For each expectation, use the stones remaining (out of ten) from Step 2 to score the reasons for the practice gap given in Step 3.
	For example, if one of the expectations was 'If I were a donkey, I would expect good quality food', four stones out of ten may be scored for current fulfilment of this expectation. The remaining stones are then divided among the reasons for the practice gap: one stone for 'not aware', three stones for 'habit/carelessness' and two stones for 'no resources'.
	You will find that coming to an agreement on this scoring will raise a lot of debate among the group.
Step 5	When scoring of all circles is complete, ask the group to identify the most important factors responsible for the welfare of their working animals. Analyse possible options for improving welfare by addressing some of the practice gaps, either individually or collectively. Issues which the group agree to act upon may be taken forward for further analysis using the Animal welfare cause and effect analysis tool (T26) and the Community action planning process (see Chapter 4, Phase 4).

Facilitator's notes: Animal welfare practice gap analysis

- This exercise can take considerable time, so discuss this in advance with the group and agree on a suitable time to set aside for doing it.
- Encourage every one to express their own views and avoid bringing in your own examples. In our experience this tool has a high risk of introducing the facilitator's reasons for practice gaps, rather than focusing on the community's reasons. Take care to avoid it becoming a facilitator-driven exercise.

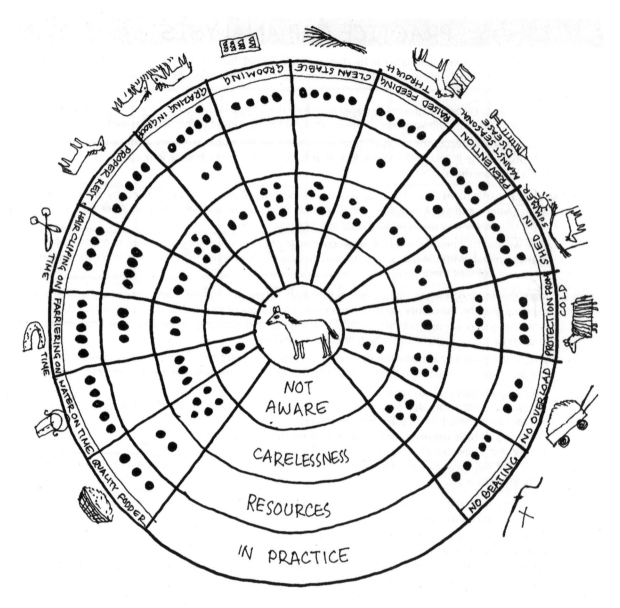

Figure T21a Animal welfare practice gap analysis carried out by donkey owners in Sona-arjunpur village, Saharanpur, Uttar Pradesh, India (2007)

A group of donkey owners from Sona-arjunpur, Saharanpur, Uttar Pradesh, analysed their current animal management practices. They identified fourteen practices that their animal would expect from them and scored the extent to which they currently carry out these practices. The highest scores were for provision of shade (a shed) in the summer and taking animals to graze in a group (six out of ten each). The lowest scores were for good quality fodder and not overloading (three out of ten each). For two practices, lack of awareness was identified as a contributing factor to gaps in implementation, and for eight of the fourteen good management practices there were resource constraints. However, carelessness (not being in the habit of doing it) was identified as a contributing reason for gaps in all the management practices. The group agreed to make these good practices a daily habit and to monitor each others' progress against their individual action plans.

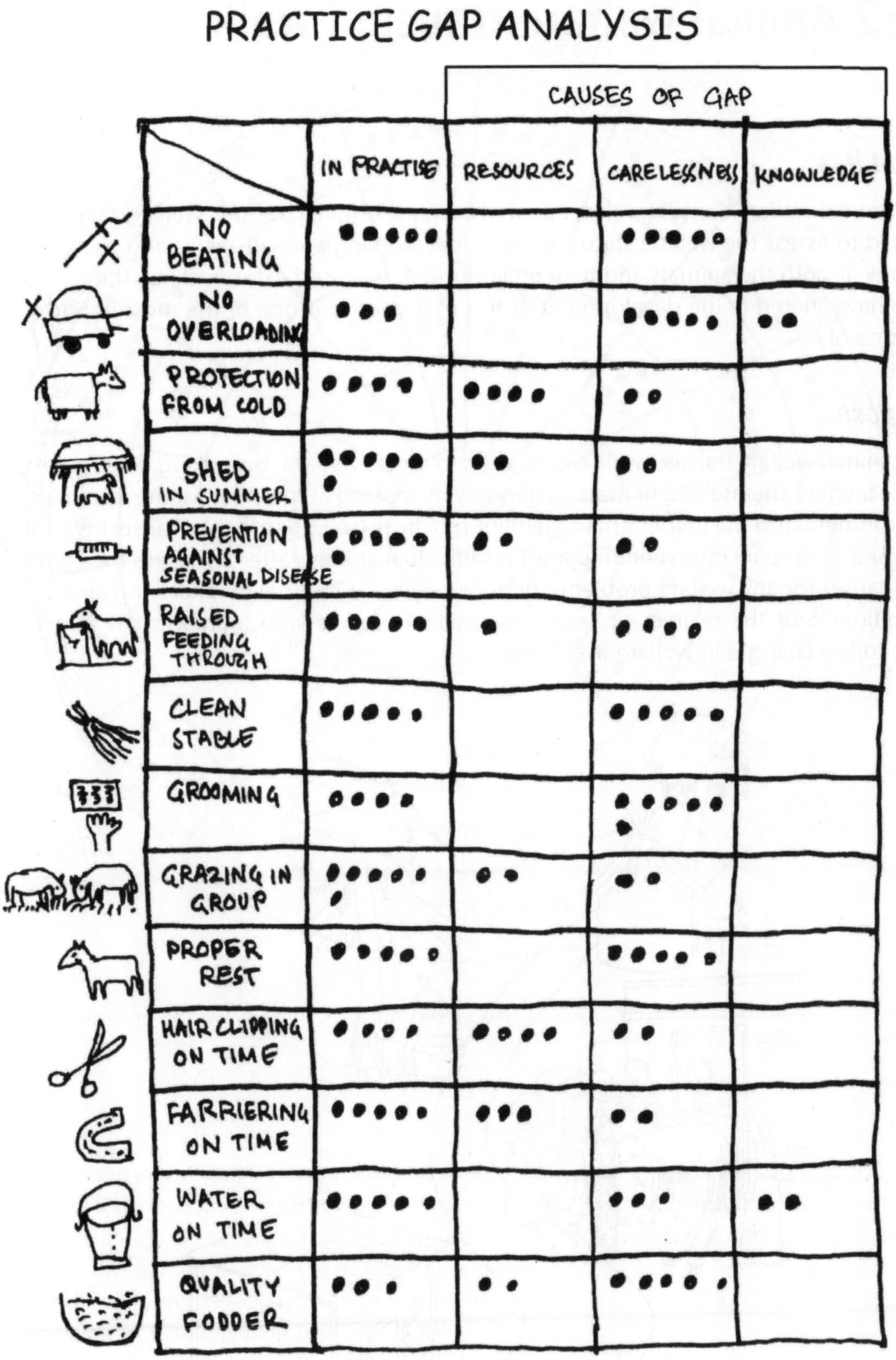

PRACTICE GAP ANALYSIS

	IN PRACTICE	CAUSES OF GAP		
		RESOURCES	CARELESSNESS	KNOWLEDGE
NO BEATING	•••••		•••••	
NO OVERLOADING	•••		•••••	••
PROTECTION FROM COLD	••••	••••	••	
SHED IN SUMMER	••••••	••	••	
PREVENTION AGAINST SEASONAL DISEASE	••••••	••	••	
RAISED FEEDING THROUGH	•••••	•	••••	
CLEAN STABLE	•••••		•••••	
GROOMING	••••		••••••	
GRAZING IN GROUP	••••••	••	••	
PROPER REST	•••••		•••••	
HAIR CLIPPING ON TIME	••••	••••	••	
FARRIERING ON TIME	•••••	•••	••	
WATER ON TIME	•••••		•••	••
QUALITY FODDER	•••	••	•••••	

Figure T21b Animal welfare practice gap analysis matrix, Saharanpur, Uttar Pradesh, India (2007)

This figure shows a similar practice gap analysis for animal welfare practices represented as a matrix rather than a circle. The exercise is carried out in the same way as described above.

T22 Animal welfare transect walk

What it is

The Animal welfare transect walk is adapted from the original transect walk (Kumar, 2002) and is used to assess the welfare status of individual working animals by visiting each household to look at both the animals and their environment. It is used extensively by the communities who have shared in the development of this manual and is one of the most important tools of our work.

Purpose

The Animal welfare transect walk can be used to assess welfare, by making direct observations of the animals themselves, or more commonly by looking at the animals, the resources in their environment and the owner's management practices (see Chapter 1). Transect walk findings are used to prepare intervention plans for individual animals after analysing the contributing root causes for any welfare problems identified. This is a key tool for exploring animal welfare conditions and the realities of resources available to working animals. It is also used for monitoring changes in welfare over time.

An Animal welfare map (T1) gives an overall 'bird's-eye' view of the animal conditions in a community, as described by their owners without the animals present. The Animal welfare transect walk gives a more complete and detailed view of animal welfare, with the animals present for the group to check and agree on, so it strengthens or triangulates the information from previous mapping exercises.

How you do it	
Step 1	Explain the purpose of the Animal welfare transect walk to the group and involve all participants in the process of decision-making about which indicators of good and poor welfare are to be observed. The observations of animals, resources and management practices may be developed from use of one or more of the following tools: • Animal welfare mapping (T1); • 'If I were a horse' (T17); • How to increase the value of my animal (T18); • Animal body mapping (T20); • Animal welfare practice gap analysis (T21); • Animal welfare monitoring criteria developed for individual or collective community action plans (see Chapter 4, Phase 4).
Step 2	Once the group has decided which observations to make, agree how the observations and discussions will be captured or recorded and who will take responsibility for this. The Animal welfare transect walk is very useful for regular monitoring of animal welfare status, as well as for action planning, so the records should be readily available to the community in the future. Decide what symbols will be used to record the result of each observation. Examples from our experience include: • Traffic light signals: good/best animal condition, resource or management practice is shown with a green dot, moderate/medium by a yellow dot and bad/worst by a red dot (see Figure T22a). Sometimes just red and green are used. • A tick for good and a cross for bad. • Symbols or colours to represent a specific problem, as agreed by the group. For example when recording whether the animal shelter is good or not, a dot of one colour is used for lack of shade, a different colour for a dirty floor and a third colour for a rough or uncomfortable floor. • Scoring each observation using numbers (see Figure T22b)
Step 3	Decide whose animals will be visited and the route of the transect walk. Fix a time when the group will go on the walk. It is essential that the owner and family of each animal visited are present when the group is visiting his or her animal and the surroundings.
Step 4	On the agreed day, the group should walk the route together, visiting all the households where animals are kept and the surrounding areas to look at each animal carefully. Discuss the agreed animal welfare indicators thoroughly with all members of the group as well as the animal owners and carers from each household visited and agree a score for each one. The group will often wish to make additions and alterations to the agreed observations. Any animal welfare issues which have emerged from other PRA tools and exercises may also be discussed with the group during the transect walk.
Step 5	After returning from the walk, ask the group to summarize and analyse their record sheets (see Figure T21a): • Adding up the scores or numbers of red, yellow and green dots vertically will give a summary of the welfare problems of individual animals. This can be used by the group to formulate individual action plans for their owners. • Adding up the scores or dots horizontally will give a score for the whole village or group for that welfare parameter. This can be used to formulate a collective action plan to reduce that welfare problem in the village. Come to a consensus on the animal welfare successes and problems and discuss possible contributing factors. Find out together whether there are workable opportunities to improve the current situation. These opportunities can be put into an action plan directly, or their causes and possible solutions may be analysed further using the Problem horse tool (T25).

TRAFFIC LIGHT

○-GOOD ◉-MEDIUM ●-BAD

Body Parts	Indicators	Umar	Islam	Zahi	Emad	Awad	Sayed	Mena	Walid
LEGS	Twisted Hoof	○	●	○	○	○	○	●	○
	Swelling Hind	◉	○	○	○	○	○	●	●
	Foot Canker	○	●	○	○	●	○	●	●
	Injury / Wound	●	○	◉	◉	●	○	●	●
	Lameness	○	●	○	○	●	○	●	○
	Stiff Legs	○	○	○	○	●	○	●	◉
EYES	One Eyed	○	○	○	○	○	○	○	○
	Whitness of Eye	○	○	○	○	○	○	○	◉
	Tears	◉	○	○	○	○	○	○	○
	Wound	○	●	○	○	○	○	○	●
EARS	Cut or Broken	○	●	○	○	○	○	○	●
	Dumb	○	○	○	○	○	○	○	●
	Droppings From Ears	○	○	○	○	○	○	○	○
	Fever	○	○	○	○	○	○	○	○
MOUTH	Teeth for Age Determination	○	○	○	○	○	○	○	○
	Flat tongue	○	●	○	●	○	○	○	○
	Like Snake	○	○	○	○	○	○	○	○
	Cut	◉	○	○	○	○	○	○	●
	Suffocation	○	●	◉	○	○	○	○	●
	Hair round on forehead	○	◉	◉	○	○	○	●	○
BACK	Wound	○	●	○	○	●	○	●	●
	Equal back	○	○	○	○	○	○	●	○
	Broken bone	○	○	○	○	○	○	○	○
TAIL	Wound	○	○	○	○	○	○	○	○
	Fatty	●	●	○	○	○	○	●	●
	Maggots	○	●	○	○	○	○	●	○
STOMACH	Ribs	○	◉	●	○	●	○	○	○
	Wound	◉	●	○	○	○	○	●	●
	Big belly	○	◉	○	○	○	○	○	○
	Low Fat	○	○	○	○	○	○	○	○

Figure T22a Animal welfare transect walk recording sheet using traffic light signals, Unnao, Uttar Pradesh, India (2009)

A horse-owning community in Unnao, Uttar Pradesh, carried out an Animal welfare transect walk using traffic lights to score 30 animal welfare indicators that they had identified during their previous discussions and exercises. After the walk the owners sat down together and analysed the score for each individual animal by summarizing the vertical columns on their recording sheet. Mena's horse was found to be in the worst condition, with 12 red ('bad') marks, followed by Islam's horse which had 11 red and two orange ('medium') welfare issues. After looking at all the individual animals, the group then summarized the horizontal rows to find out which welfare issues were most common in their village. Leg injuries and wounds were the biggest problem in their community, with four animals marked red and two more marked orange. The group followed this exercise with a Root cause analysis (see Chapter 4, Phase 4 and T25 and T26) and a community action plan for immediate action by individual owners and collectively.

status / Owners	Equine Type	Status of Animal	Firing	Wound	Body cleaning	Hair clipping	Eye/Mouth cleaning	Stable cleaning	Manger cleaning	Water availibility	Hobbling	Foot canker	TOTAL
Mahendra	Mule	10	10	10	5	10	5	10	10	5	5	10	90
Puran	Mule	5	10	5	5	10	5	10	10	8	5	10	83
Prahalad	Horse	5	10	5	5	10	5	10	10	8	5	5	78
Kallu	Horse	5	10	5	5	10	5	10	5	5	5	5	70
Bablu	Horse	5	10	10	8	5	5	10	10	7	5	10	85
Netar	Mule	0	10	0	5	10	5	10	5	0	5	5	55
Dharampal	Horse	5	10	6	10	10	10	5	5	5	5	10	81
Ravi	Mule	5	10	7	5	10	5	10	2	5	5	10	74
Pappu	Mule	10	10	5	5	10	5	10	10	10	5	10	90
Sher Singh	Horse	5	10	8	0	8	5	10	10	8	5	5	74
Ranveer	Mule	10	10	10	10	10	10	10	8	18	5	10	103

Figure T22b Animal welfare transect walk recording sheet using numerical scoring, India (May 2008)

The Dhanoura village community group (Bulandsahar, Uttar Pradesh) went on an Animal welfare transect walk and scored 11 animal welfare indicators out of a maximum of 10 marks each. Ranveer's mule was in the best condition, scoring 103 out of 110 overall, and Netar's mule was scored the lowest, at 55 out of 110. They also identified the overall status of animals in the village by adding up the scores for each column (not shown in this illustration), with the lowest scores representing the most common welfare problems. The lowest scoring problem overall was hobbling (scoring 55) followed by cleanliness of body (63), body condition (65) and eye and mouth cleanliness (65). After this exercise the group carried out a root cause analysis of the four animal welfare problems with the lowest score using the Animal welfare cause and effect analysis (T26) and went on to develop a community action plan for animal welfare improvement (see Chapter 4, Phase 4).

Facilitator's notes: Animal welfare transect walk

- Transect walks look in detail at the group's real experiences of animal welfare conditions, the existing use and availability of resources and the animal management constraints being faced by the whole community and by individual animal owners.
- Usually we meet on one day to agree all the observations and processes for walking and recording. The transect walk is carried out on another pre-determined day to suit the community.
- If there are large numbers of animals, carry out the same exercise over several days so that all animals or a representative number of animals and households are covered.
- Animal welfare transect walks may be repeated at regular intervals and the results compared with previous walks. This enables the group to monitor and evaluate changes in the welfare status of individual animals, changes in their owners' management practices and changes in the animal-related resources available.

T23 Three pile sorting

What it is

With Three pile sorting, cards are used to enable a community group to sort and discuss animal management and work practices according to whether they are seen as good, bad or neutral for animal welfare. It is adapted from a previous card sorting tool (Rietbergen-McCracken and Narayan-Parker, 2006).

Purpose

Three pile sorting is used to explore participants' understanding and perspectives on any animal welfare issue and to provide a starting point for problem analysis and action. For example, you might ask which management issues are commonly found in the community, whether they are good or bad and what might be done to reduce the effect of poor welfare practices. Issues such as putting animals to work at an early age, firing or branding and overloading may be addressed using this tool. Where management practices are seen as neither good nor bad, the group might discuss why this is so. We have also used it to analyse perceptions about animal diseases and their symptoms, causes and prevention.

How you do it	
Step 1	For this exercise you need to prepare beforehand.
	Make a set of set of cards showing animal welfare or management practices which can be interpreted as good, bad, or in-between (neutral). These should be based on problems previously identified by the community during exercises such as 'If I were a horse' (T17), Animal body mapping (T20) or Animal welfare practice gap analysis (T21). Common negative practices which could be illustrated on the cards include beating animals, overloading, not offering water, incorrect feeding and lack of care for wounds.
Step 2	Organize participants into groups of no more than seven and ask each group to form a circle. Give a set of cards to each circle and ask for two or three volunteers to sit in the centre of the circle and sort the cards together. They should place each card in one of three piles: representing good welfare or management practices, bad welfare or management practices, and in-between or neutral practices (or practices where there is uncertainty or disagreement).
Step 3	After the cards have been sorted, ask the volunteers to arrange them so that each card is visible to the whole group. Encourage debate between participants to challenge their choices and analyse all aspects of their decision. Enable participants to understand why a particular practice may be better or worse for animal welfare.
Step 4	Ask the group to identify which of the animal welfare issues or management practices are occurring in their own village, especially the ones they have identified as bad. This discussion can be used to enable participants to identify priority welfare problems and to propose potential solutions or action to be taken.

Facilitator's notes: Three pile sorting

- It is important that you to enable the group to bring their own perceptions and use their local terms for management practices, diseases, pain and animal suffering.
- The game can also be played using photographs representing the existing situations in the village. This needs advance preparation.
- Another variation is to ask participants to sort cards according to different types of animal welfare issue they have experienced, such as i) issues directly related to animals ii) issues related to service providers and other stakeholders and iii) issues related to animal owners, users and carers.

T24 Animal welfare story with a gap

What it is

This tool uses a 'before and after' story to stimulate discussion about how to change from a situation of poor animal welfare to a situation where welfare is improved. It is adapted from a similar tool (Rietbergen-McCracken and Narayan-Parker, 1996).

Purpose

Animal welfare story with a gap uses a pair of pictures, one showing a 'before' situation relating to a working animal and the other showing an 'after' scenario where the animal's welfare has improved. For example, the 'before' drawing might show an animal being beaten to make it work, and the 'after' could have the same animal being worked using only voice commands. Participants discuss both drawings and fill in the gap in the story by identifying the steps that would need to be taken to achieve what is represented in the improved picture.

How you do it	
Step 1	For this exercise you need to prepare the pictures beforehand: use drawings or photographs of existing animal management situations or practices in the community.
Step 2	Divide the participants into several small groups and give each group the same set of 'before' and 'after' pictures. Ask each group to begin by considering the 'before' picture, such as a picture of a working animal with wounds, and to discuss why the situation has occurred. Next, ask each group to discuss the 'after' scene of the improved situation, such as an animal with fewer or no wounds. Then ask the groups what steps they think they might take to get from the 'before' to the 'after' scenario (in other words how they would fill the gap in the story), what obstacles they might have in their way, and what resources they would need to do this.
Step 3	Bring the different groups together and ask each group to tell the stories they have created. Encourage the groups to weigh the benefits of each suggestion for welfare improvement and discuss more ways to overcome the obstacles.

Facilitator's notes: Animal welfare story with a gap

- Your facilitation of the discussion of 'before' and 'after' scenarios should ensure that the animal welfare context is clearly shown.
- More interpretations and suggestions can be gathered by dividing the participants into several small focus groups (for example of women and men, young and old people, or other categories) and giving each the same set of pictures. After analysing the drawings, the focus groups can come together to report on their discussions and compare their views.

T25 Problem horse

… or donkey, mule, bullock, camel or yak

What it is

Problem horse is a tool that we have adapted from the Cause and effect diagram (T26) in order to carry out a Root cause analysis of animal welfare problems. The original version is described in *Methods for Community Participation* by Somesh Kumar (see the further reading and reference list).

Purpose

The Problem horse exercise (Figure T25) enables a group to identify the welfare issues affecting different parts of an animal's body and to recognize the relationship between each welfare issue and its possible causes. The first step is the same as Animal welfare body mapping (T20).

How you do it	
Step 1	Ask participants to draw a large animal shape on the ground or on a piece of chart paper. Help them to identify welfare issues related to each part of the body and draw them on the picture of the animal. In the illustrated example below (Figure T25) the group identified watering eyes, colic and wounds on the hind leg, belly, girth and under the tail.
Step 2	Choose any one of the welfare issues identified in Step 1. Encourage participants to discuss the causes of the problem and draw or write the causes near the relevant part of the animal's body. Analyse each welfare issue in depth by repeatedly asking 'why?' questions. For example:
	'Why does the animal get that wound?' – 'Because of the leather belt on the harness'
	'Why does the leather belt cause the wound?' – 'Because it is not cleaned and oiled'
	'Why is the belt not cleaned and oiled?' – 'Because we don't have the time'
	'Why don't you have the time?'
	and so on, until the group reaches the deepest root causes of the welfare problem and cannot go any further.
	When one welfare problem is complete, take up the next one and repeat the questions until root causes are drawn or written next to all the problems shown on the body of the animal.
Step 3	As the discussion progresses and all the causes are identified, analyse any links or relationships between different causes and show these using lines or arrows (see Figure T25).
Step 4	Enable the group to identify the most important causes and use an exercise such as Matrix ranking (T9) to prioritize them.
Step 5	Discuss what action the group can take to tackle the causes of each priority welfare issue. If they are unable to remove the cause, ask whether there are measures they can take to reduce its effect on their animals. Encourage them to draw up a community action plan to deal with these (see Chapter 4, Phase 4).

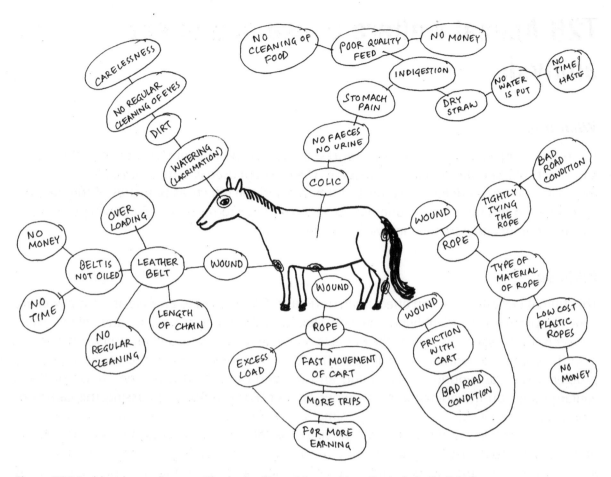

Figure T25 Problem horse diagram, Kharkoda village, Meerut, Uttar Pradesh, India (2007)

This diagram was made by a group of animal owners from Kharkoda village, Meerut, Uttar Pradesh. First, the group discussed the problems affecting each part of the animal's body, then they analysed the problems in depth to find their root causes. The group found that wounds on different parts of the body have different causes, but there are also causal factors which are common to more than one body area, such as bad road conditions and the way that ropes are tied. Wounds at the back and wounds on the chest were found to be inter-related. The group drew up collective plans for action to address some of the root causes identified during this exercise.

Facilitator's notes: Problem horse

- In India we have used a 'broken horse' jigsaw puzzle as the starting point for this exercise. We initiate discussion about the body of the animal using a wooden jigsaw of animal body parts, which owners put back together to make a complete picture of the animal.
- This exercise needs a lot of patience and questioning in order to enable the deepest causal factors to come out of the discussion.
- We sometimes find that owners return to the Problem horse (T25) tool (or other root cause analysis tools such as the Animal welfare cause and effect analysis T26) when they have implemented an action plan and it has not succeeded in creating the expected welfare improvement. This leads the group to reflect on their analysis and ask whether they have missed any important root causes (see Chapter 4, Phase 5).

T26 Animal welfare cause and effect analysis

What it is

A cause and effect analysis is a visual representation of the relationship between the causes and effects of a specific animal welfare issue. The diagram used is sometimes called a problem tree, with the causes depicted as roots of the tree and the effects as branches of the tree. We have adapted this tool from its original use (Kumar 2002) to represent relationships between the community and particular individuals or institutions.

Purpose

The Animal welfare cause and effect analysis (Figure T26) is used to analyse an animal welfare issue or problem by identifying the complex contributing factors and any relationships between these factors, as well as their effects on the animal and its owner, users and carers.

In the context of working animals we often use this tool to identify the causes of welfare problems such as wounds, overloading and hobbling and to discuss the effects of these welfare issues on the animal and its owner or user. For example, discussing the causes of wounds on specific parts of a working animal's body may highlight causal factors such as the structure or cleanliness of the harness, the size of saddle tree or the design of the cart. Effects on the animal could include pain, weight loss and reduced working capacity. Effects of these animal wounds on the owner could include less income (from reduced work and increased expenditure on medicines) or lower status in the community.

How you do it	
Step 1	Ask the group to list the animal problems or issues that they have identified using other exercises, such as Animal body mapping (T20), Animal welfare practice gap analysis (T21) or the Animal welfare transect walk (T22). If there are too many problems to analyse in one meeting, rank them using Pair-wise ranking (T8) or Matrix scoring (T9) to agree on the most important item(s) for an Animal welfare cause and effect discussion.
Step 2	Draw a circle on the ground or on paper, putting one welfare problem in the centre. Ask the group about the major factors which cause this particular problem. Identify the major causal factors and show them outside the circle using symbols, pictures or words. Connect these to the problem with arrows.
Step 3	Take up one particular causal factor and ask participants why this happens. The group will start to discuss the sub-causes which contribute to this major cause. Show these sub-causes outside the major cause, connecting them to the major cause with arrows (see Figure T26). For each sub-cause, again ask the question 'And why does this happen?' Keep asking the question and adding sub-causes, branching out like the roots of tree, until the group reaches a stage where no further answers can be found.
Step 4	When all causal factors are analysed and discussed, ask participants to identify the effect of the welfare problem on the animal and its owner. Draw two circles, one representing the animal and one for the owner, and connect these with the problem identified (see Figure T26). Draw the effects around the animal and owner circles, discussing each in depth.
Step 5	Work out which are the most important causal factors contributing towards this animal problem through discussion or by using Pair-wise ranking (T8) or Matrix ranking (T9). Encourage the group to develop a community action plan to tackle these causal factors and improve their chosen welfare problem.

Based on their analysis, the community can decide on appropriate action to improve the welfare problem and develop a community action plan containing both individual and collective actions. This tool also helps to sensitise and motivate the community to act on welfare issues which were not initially recognized as important for animals.

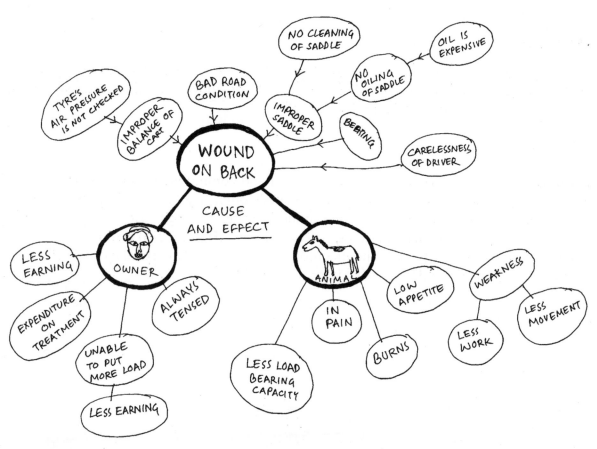

Figure T26 Animal welfare cause and effect diagram for wounds on the back, Ghaziabad, Uttar Pradesh, India (2007)

This diagram is the output of an Animal welfare cause and effect analysis by a group of horse owners working in a brick kiln in Ghaziabad, Uttar Pradesh. They were particularly concerned about reducing wounds on their animals' backs. Five major causes were initially identified: poor cart balance, bad road conditions, an improper saddle, beating and careless driving. These were analysed further to reach to the root causes. The effects of back wounds on the horses were seen as weakness, low appetite, pain and reduced ability to carry loads. The effects on the owners were found to be lower income, higher expenditure on treatment and always feeling tense. The group went on to take action on the root causes which were within their influence, by checking tyre pressures on the brick carts, driving more carefully and cleaning their saddles regularly.

Facilitator's notes: Animal welfare cause and effect analysis

- This exercise can take considerable time, so discuss this in advance with the group and agree on a suitable time to set aside for doing it.
- Avoid using your own examples and encourage everyone to express their individual views. Allow enough time for participants to discuss their experiences.
- This exercise needs lot of patience in order to facilitate the deepest causal factors to merge from the discussion.

T27 Analysis of animal feeding practices

What it is

The Animal feeding practice analysis tool is used to study feeding practices in depth. The analysis integrates the community's knowledge with the knowledge of an external expert, such as a veterinarian or animal nutrition specialist.

Purpose

In our experience many working animals are thin and community groups identify feeding as one of their major concerns. Root cause analysis often highlights difficulties in finding enough animal food of good quality and nutritional value; the high cost of animal feed and fodder is also a concern. Animal feeding practice analysis (Figure T27a) enables participants to look closely at their existing feeding practices and develop a low-cost balanced diet for their animals, based on feed and fodder that is locally available.

How you do it

This exercise needs a veterinarian or animal nutritionist to join the group during Step 2, which should be arranged in advance.

Step 1	*Identifying current animal feeding practices* Seasonal analysis (T6) can be used as the basis for identifying current feeding practices. Ask group members to list the types of feed they give to their working animals (using local names) and represent the quantity fed in a month using stones or seeds. Then add a new row to identify the cost of animal feed per day in each season. Alternatively feed costs could be listed in a separate grid, as shown in Figure 27a, Illustration 1.
Step 2	*Identifying the nutritional value of each type of feed* Develop a table or a circular diagram as shown in Figure 27a, Illustration 1. Enable the group to identify the purpose of different animal feeds by asking a question such as: 'What does this feed do in the animal's body?' Answers such as 'power for work', 'helps good digestion' and 'for filling the belly' are drawn or written on the diagram. Invite the veterinarian or animal nutritionist to join the group at this stage and provide technical input on balanced feeding using locally available feed ingredients (Figure T27a, Illustration 2).
Step 3	*Identifying the feed requirements of working animals* Ask the group to discuss whether they should provide equal quantities of a specific feed to all their animals or whether different animals need different amounts of each feed. Ask whether the amount that they feed their working animals is the same in all seasons or if it varies between the peak working season and the off-season. With the help of the veterinarian or animal nutritionist, develop a table showing feeding requirements for working animals, including factors such as body weight and work load.
Step 4	*Developing options for balanced feeding* Based on the requirements of their individual animals, the nutritional value of local ingredients and the feed costs, enable the group to search for several options for feed mixtures that are available, affordable and have appropriately balanced ingredients.
Step 5	*Developing individual action plans* Ask the group to decide on one of the feed mixtures for each animal, according to its owner's preference, and to experiment with feeding the new balanced feed mixture for a few months (see Figure T27a, Illustration 3).
Step 6	*Monitoring the effects of the new diet* Ask group members to decide on the indicators they will use to judge whether the new diet is successful; for example reduced visibility of the ribs, improved thickness in the neck and back region and not getting tired easily. Owners should provide feedback during their monthly meetings and discuss their findings, adjusting each animal's diet according to its effect on the indicators that they have chosen.

Facilitator's notes: Analysis of animal feeding practices

- This exercise can take considerable time, so you should discuss this in advance with the group and agree on a suitable time to set aside for doing it. Sometimes we divide it over two sessions to include all feeds and aspects of feeding practice.
- Facilitate the group to recall all the types of feed that they use in a year. We find that at first they only include the feeds which they are giving in the present season.
- Ensure that the veterinarian or animal nutritionist is sensitive to the constraints affecting animal owners and knows about local feeds which are not too expensive, complicated or difficult to obtain.

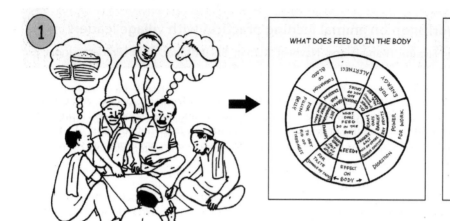

WHAT DOES FEED DO IN THE BODY

FEED WE ARE GIVING TO OUR ANIMAL

WHAT (FEED)	AMOUNT	TOTAL EXPENSE (RUPEES)
WHEAT STRAW	4 KG	20
CHURRIE (SORGHUM SUDANESE OR MAZE OR MIX OF BOTH)	2 KG	24
WHEAT BRAN	2 KG	22
GREEN GRASS	20 KG	20
MIX OF SPICES		15
TOTAL COST		101

TECHNICAL PRESENTATION BY VET

INGREDIENT	SUPPLY ENERGY HIGH	SUPPLY ENERGY LOW	MAINTAIN BODY HIGH	MAINTAIN BODY LOW	FILLING STOMACH HIGH	FILLING STOMACH LOW
WHEAT STRAW		✓		✓	✓	
PADDY STRAW		✓		✓	✓	
RICE BRAN	✓			✓	✓	
BARLEY	✓		✓			✓
OAT	✓		✓			✓
GRAM	✓		✓			✓
MOLASSES	✓			✓		✓
CLARIFIED BUTTER	✓			✓		✓
MUSTERED OIL	✓			✓		✓
MILK		✓	✓			✓
BREAD		✓		✓		✓

ACTION PLAN

INGREDIENT	AMOUNT	COST IN RUPEES
WHEAT STRAW	3 KG	15
WHEAT BRAN	2 KG	22
GRAM	1 KG	25
JAGGERY	0·5 KG	11
GREEN GRASS	15 KG	15
OIL / SALT	100gm	07
TOTAL	21·600 kg	95

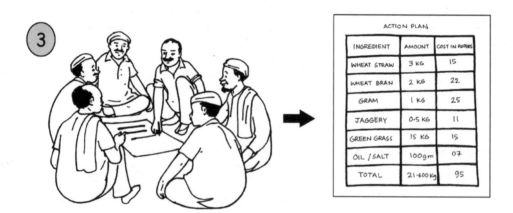

Figure T27a Process steps in the analysis of animal feeding practices

Case study O. Workshop on animal feeding practices with village leaders

Source: Kul Bahadur Fagami (Kamal) and Navneet Kumar, Bulandsahar, Uttar Pradesh, India, January 2009

In January 2009, 26 participants from 14 working equine welfare groups attended a workshop for animal welfare group leaders in Bulandsahar. The leaders had asked for the workshop in order to analyse their current animal feeding practices and generate a cheap and balanced feed for working horses, mules and donkeys. The session started with a discussion on the topic: what is needed to ensure working animals have a happy and healthy life? Group members listed food, water, shelter, clothing (blanket), exercise and rest in between working periods.

This was followed by the question 'What are the things which most affect the life of an animal?' The group listed:
- feed and fodder;
- shelter;
- water.

Next they were asked: 'Which one is most expensive?' The group responded:
- water is available free of cost;
- shelter requires a one-time investment;
- feed and fodder require a daily investment, so they are most expensive.

The fourth question was *'Why do we feed our animals and why is the food important?'* which resulted in a list of what food does in the body: supplying power, maintaining the body and filling the stomach. Participants were divided into groups and asked to sort all the feeds they used during the year into these three categories. At this stage a veterinarian joined the meeting and presented technical nutritional information about the feeds listed. The group compared the vet's presentation with their own lists and some discussion followed. With more help from the vet, participants determined the concentrate and fodder requirements for animals of different body weight and with different work loads.

In the last session of the workshop, the group discussed the basic principles of good feeding practices and were placed in groups again to formulate an improved feed mixture according to their new and shared knowledge. This was more balanced and cheaper than the mixtures that they were feeding originally (see Figure T27b).

FEED INGREDIENTS	PURPOSE	QUANTITY (kg)	COST (kg) (RUPEES)	TOTAL COST (RUPEES)
GRAM	Protein	0.50	30	15
CHANACHUNI	Protein	0.75	13	10
BARLEY	Energy	0.50	7	3.5
OAT	Energy	1.50	20	30
WHEAT BRAN	Energy + Bulk	1.00	10	10
CURD	Energy	0.25	20	5
VEGETABLE OIL	Energy	0.50	60	3
SALT		0.50	3	1.5
TOTAL		5.50		78

Figure T27b Balanced diet for working animals formulated by a group of village leaders, India (January 2009)

Several group leaders were very enthusiastic but others were a bit sceptical and wanted to try the new diet on their own animals before involving other members of their group. At the end of the day it was agreed that they all would offer the new feed mixture to their own animals for two weeks. If it was successful they would conduct a meeting to introduce it to group members.

The feedback received by the workshop organizers after two weeks was very positive. Twenty group leaders had tried the new feed and saw improvement in their animals. Several went to visit other group leaders to look at their animals and see if the new balance was successful. Most of these group leaders have started to sensitize their own group members to the new feed and have requested support from the vet to provide more technical input for individual animals.

T28 Village animal health planning

What it is

The Village animal health planning tool uses a combination of the tools in this toolkit to explore and address specific health problems affecting working animals.

Purpose

Village animal health planning is a flexible process which uses several tools to enable the group to gain a deeper understanding of animal health problems, including patterns of disease occurrence, causes, symptoms, treatments, disease-related expenses, recovery rates and preventive measures. This leads to a community action plan for improving animal welfare by preventing or reducing the effects of common injuries or diseases. If the village is affected by a disease outbreak or a specific animal health issue, health planning may be a good starting point for other welfare-promoting activities.

How you do it

Process box 10 below lists the tools which we use for the Village animal health planning, in the order we usually use them. However they can be carried out in any sequence based on your judgement and the particular needs of the community. You will see that we have suggested focusing some of these tools more closely on health issues than you would do if using them for more general animal welfare improvement purposes. Case study P below illustrates a real example of the process. Our field teams have made many adaptations to this process to make it fit the specific needs of the communities that they work with.

This is only a short introduction to the benefits and process of the Village animal health planning tool. A more detailed booklet will be published in the future.

Process box 10. Tools for village animal health planning

Historical timeline (T7)
Historical events related to working animals in the village or wider locality and the diseases or epidemics that have affected them.

Changing trend analysis (T11)
Analysis of changes in the working animal population, feeding practices, nature of their work, occurrence of diseases, mortality rates, treatment practices, availability of animal health services and medicines, treatment costs, design of carts and harnesses.

Animal disease mapping (T1)
Mapping of households, including the number of working animals in each household, present health status, past occurrence of disease (for example during the last two years), location of veterinary services and medicine shops.

Animal welfare transect walk (T22)
The group discusses disease or health problems with their potential causes and possible sources of contamination or spread. Then participants visit each animal and identify any signs of disease on each animal and its causes in the environment. A discussion on the immediate and long term actions needed for prevention often follows straight away.

Seasonal analysis (T6)
Analysis of the occurrence of various diseases in particular seasons or months of the year.

Matrix ranking (T9)
Ranking of diseases and injuries according to their importance, in terms of their severity and frequency of occurrence.

Matrix scoring (T9)
Scoring of animal diseases and injuries against their common symptoms, causes, numbers of animals affected, mortality, severity, recovery rates, loss of working time, loss of working efficiency and loss of income or increased expenditure.

Problem horse (T25) or Animal welfare cause and effect analysis (T26)
Analysis of the causes and effects of a specific disease or health problem prioritised by Matrix ranking or Matrix scoring.

Analysis of present animal treatment practices
What do we do when our working animal gets ill? Use the grid below (or a similar one according to participants' choice) to discuss who treats which disease, recovery times, disease-related expenses, number of cases and the success rate of treatment.

DISEASE	TREATED BY	NUMBER OF DAYS TO RECOVER	TOTAL EXPENSE	NUMBER OF CASES TREATED IN LAST 1-5 YEARS	SUCCESS RATE
TRYPANOSOMIASIS					
TETANUS					
LAMENESS					
WOUNDS					
COLIC					

Figure T28a Village animal health planning: matrix of present treatment practices

Village action planning
Development of an action plan for immediate and long term disease prevention and treatment (Chapter 4, Phase 4).

Case study P. Village animal health planning

Source: Murad Ali, Brooke India, Jaisinghpur, Meerut, India, September 2009

Mukesh Kumar, an animal welfare facilitator in Meerut district, visited Khandawali village in September 2008. One owner brought his horse with the complaint that it was sick. He had consulted other animal owners but they did not know what it was suffering from. Mukesh called the veterinarian and the horse was diagnosed as suffering from equine influenza, a highly contagious disease. All the village horse owners realized that other animals were also affected so they held a meeting to discuss the disease.

They started by drawing a map of the village and used symbols and arrows to show the first animals with influenza and the way it had spread, such as by direct contact with diseased animals in vegetable markets, drinking water in a communal water trough and grazing with affected animals in village. The owners categorised the severity of each animal's infection using Matrix scoring of their disease signs, such as looking dull and depressed, having fever, a discharge from the nose and a dry cough.

There were still some animals in the village which were not showing signs of influenza. During the discussion it was found that those animals either did not work, or did not graze with other animals. A list of preventive actions was developed for all owners to follow, including use of individual water buckets, cleaning of stables and keeping diseased horses away from other horses while grazing. Several owners took responsibility for monitoring implementation of these preventive actions every second day. The group agreed to meet once a week to monitor whether fewer animals were showing signs of influenza.

During a later visit to Khandawali village, the animal owners told Mukesh that they felt particularly vulnerable to animal diseases and asked him to work with them to prevent disease, improve their treatment practices and in this way save money as well. During the day the horse owners were busy working at the brick kiln, so Mukesh agreed to meet them every evening for two to three days in order to start the process of Village animal health planning. The first meeting started with a disease-related Historical timeline to identify which animal diseases occurred in Khandawali during the last five years. The diseases were then scored for severity (the number of animals which had died in five years) and frequency (how many times the disease had happened in five years). Six diseases or health problems were listed in order of priority: surra (trypanosomiasis), tetanus, lameness, wounds, breathing problems and colic (abdominal pain).

The group of owners discussed the symptoms and causes of these diseases and health problems and realized that many of them could be prevented by good management practices. They invited local health providers to be part of this discussion and identified their own present practices for disease prevention and treatment and the practices of the local healer, medical store and private and government veterinarians. This discussion initiated an in-depth analysis of who does what, the recovery times, costs, and success rates of each treatment.

Based on this detailed situational analysis, the group developed a community action plan containing three types of intervention: short term, medium term and long term (see Figure T28b). They have used this to prevent wounds, infections and colic, identify diseases early and negotiate reduced prices for vaccination and treatment of their working horses.

	DISEASE	ACTION	MONITORING
SHORT TERM ISSUE	WOUNDS LAMENESS COLIC INFLUENZA	- Regular transect walk to monitor welfare and identification of diseases - Community medicine kits - Two members of group monitor influenza on weekly basis	• Analyse transect walk recording charts at monthly meeting • Group leader keeps record of community medicine kit • keeping treatment records in the group and analyse during monthly meeting • Initiate immediate action if required and report in monthly meeting
MEDIUM TERM ISSUES	TETANUS, RABIES and TRYPANOSOMIASIS PREVENTION	- Community led immunisation/ vaccination plan - Agree on community sanitation such as spray of lime and fly prevention	• Check vaccination records every 3 months in group meeting • See action during transect walk
LONG TERM ISSUES	Firing done by local healer High treatment cost of local doctor Availability of Government doctor	- Capacity enhancement healer request Brooke for support - Negotiate with local doctor as a group reasonable prices - Availing services from Government veterinary hospitals	• Through transect walk monitoring cases of firing

Figure T28b Village animal health action plan, India (September 2008)

References and further reading

This list of references and further reading includes materials on community participatory approaches, community outreach methods, animal welfare and working animals. References are organized according to the chapters where they are cited, with a general further reading list at the end.

Preface

Livestock Emergency Guidelines and Standards (LEGS) (2009) Practical Action Publishing, Rugby, UK. www.livestock-emergency.net/downloads.html [online].

Introduction

Main, D.C.J., Kent, J.P., Wemelsfelder, F., Ofner, E., and Tuyttens, F.A.M. (2003) 'Applications of on-farm welfare assessment', *Animal Welfare* 12, 523–528.

Chapter 2

Five Freedoms (1979) Farm Animal Welfare Council, UK www.fawc.org.uk/freedoms.htm [online].

Fraser, D., Weary, D.M., Pajor, E. A., and Milligan, B.N. (1997) 'A scientific conception of animal welfare that reflects ethical concerns', *Animal Welfare* 6, 187–205.

Webster, A.J.F, Main, D.C.J., and Whay, H.R. (2004) 'Welfare assessment: indices from clinical observation', *Animal Welfare* 13, 93–98.

Chapter 3

Catley, A., Blakeway, S., and Leyland, T. (2002) *Community-based Animal Healthcare: A Practical Guide to Improving Primary Veterinary Services*, BookPower/ITDG Publishing, Rugby, UK.

Kumar, S. (2002) *Methods for Community Participation: A Complete Guide for Practitioners*, Intermediate Technology Publications Ltd, Rugby, UK.

Pretty, J.N. (1995) *Regenerating Agriculture, Policies and Practices for Sustainability and Self Reliance*, ACTION AID, Bangalore, India.

Pritchard, J.C., Lindberg, A.C., Main, D.C.J., and Whay, H.R. (2005) 'Assessment of the welfare of working horses, mules and donkeys using health and behaviour parameters', *Preventive Veterinary Medicine* 69, 265-283.

van Dijk, L., and Pritchard, J.C. (2010) *Determinants of Working Animal Welfare*, Adapted from Dahlgren G., and Whitehead, M. (1991) *Policies and Strategies To Promote Social Equity In Health*, Institute of Future Studies, Stockholm, Sweden.

Chapter 4

Bhatia, A., Sen C.K., Pandey,G., and Amtzis, J. (eds). (1998) *Capacity-Building In Participatory Upland Watershed Planning,* Monitoring and Evaluation, a Resource Kit, ICIMOD, Kathmandu, Nepal, pp. 55–59 http://books.icimod.org/index.php/search/subject/4/45 [online].

Capeling-Alakija, S., Lopes, C., Benbouali, A., and Diallo, D. (1997) *A Participatory Evaluation Handbook - Who are the Question Makers?* OESP Handbook Series, Office of Evaluation and Strategic Planning, United Nations Development Programme, New York, USA.

Davies, R., and Dart, J. (2005) *The Most Significant Change Technique: a Guide to its Use* www.mande.co.uk/docs/MSCGuide.htm [online].

Society of Friends Peace Committee (undated) *The Two Mules, a Fable for the Nations: Co-operation Is Better Than Conflict,* Washington, USA.

Elliott,C. (1999) *Locating the Energy for Change: An Introduction to Appreciative Inquiry,* IISD www.iisd.org/publications/pub.aspx?id=287 [online].

Chapter 5

Burke (1999) *Communication and Development: A Practical Guide,* Social Development Division, Department for International Development, London, UK www.dfid.gov.uk/Documents/publications/c-d.pdf [online].

Theatre for development

Tearfund (2004) *Community Drama – Theatre for Development,* Theatre for Development, Footsteps no. 58 http://tilz.tearfund.org/Publications/Footsteps+51-60/Footsteps+58 [online].

The Royal Tropical Institute lists a number of references on theatre and development http://insightshare.org/resources/pv-handbook [online].

Tufte, T., and Mefalopulos, P. (2009) *Participatory Communication: A Practical Guide,* World Bank Paper 170 http://orecomm.net/wp-ontent/uploads/2009/10/Participatory_Communication.pdf [online].

Participatory video

Lunch, N., and Lunch, C. (2006) *Insights Into Participatory Video: A Handbook For The Field.* Insight, Oxford, UK http://insightshare.org/resources/pv-handbook [online].

Community radio

Barker, K. (2008) *Community Radio Start-Up Information Guide,* Farm Radio International, www.farmradio.org/english/partners/resources/community-radio-start-up-guide_e.pdf [online].

Ochieng, F., and Kaumbutho, P. (2006) *Extension Approaches and Radio Messaging for Improving the Welfare of Working Equines: A Manual for Implementing a Radio Programme to Improve Equine Welfare,* www.kendat.org/eLibrary/link.php?id=46 [online].

Tabing, L. (2002) *How to do Community Radio.* UNESCO Asia-Pacific Bureau For Communication and Information, New Delhi, India http://unesdoc.unesco.org/images/0013/001342/134208e.pdf [online].

Toolkit

Jayakaran, R. (2007) *Holistic Worldwide View Analyses: Understanding Communities' Realities,* PLA Notes no. 56, International Institute for Environment and Development, London, UK.

Kumar, S. (2002) *Methods for Community Participation: A Complete Guide for Practitioners,* Intermediate Technology Publications Ltd, Rugby, UK.

Rietbergen-McCracken, J. and Narayan-Parker, D. (1996) *Participation and Social Assessment: Tools and Techniques* World Bank Resource Book pp. 316–320.

Thomas-Layter, B., Polestico, R., Lee Esser, A., Taylor, O., and Mutua, E. (1995) A *Manual For Socio-Economic Gender Analysis: Responding to the Development Challenges.* Clark University, Massachusetts, USA.

The Brooke is a leading UK animal welfare charity dedicated to improving the lives of horses, donkeys and mules working in the poorest parts of the world. There are an estimated 100 million working equine animals in the world and they are often the only source of income for many poor families, who depend on their working animal to survive.

The Brooke – along with our partner organizations worldwide – supports mobile veterinary teams and community animal health workers who provide treatment to working animals and advice to their owners. Our community engagement specialists build the capacity of animal owners, local healers, farriers, saddlers, feed sellers, harness- and cart-makers to look after their animals' needs in a sustainable way. We currently operate across 10 countries in Asia, Africa, Central America and the Middle East. We have over 800 staff working directly in the field.

In 2009 we reached 730,000 working horses and donkeys – benefiting over 3.7 million poor people who depend on them. Our goal is to improve the welfare of two million animals every year by 2016.

For more information about the Brooke please visit our website: www.thebrooke.org